Polymeric Liquid Crystals and Applications

Polymeric Liquid Crystals and Applications

Guest Editors

Pedro Marques de Almeida
Ana Catarina Trindade
João Canejo

MDPI

Basel • Beijing • Wuhan • Barcelona • Belgrade • Novi Sad • Cluj • Manchester

Guest Editors

Pedro Marques de Almeida
Mechanical Engineering Department
Polytechnic University of Lisbon
Lisbon
Portugal

Ana Catarina Trindade
Atlântica—Instituto Universitário
Barcarena
Portugal

João Canejo
NOVA University Lisbon
Caparica
Portugal

Editorial Office
MDPI AG
Grosspeteranlage 5
4052 Basel, Switzerland

This is a reprint of the Special Issue, published open access by the journal *Molecules* (ISSN 1420-3049), freely accessible at: https://www.mdpi.com/journal/molecules/special_issues/Polymeric_Crystals.

For citation purposes, cite each article independently as indicated on the article page online and as indicated below:

Lastname, A.A.; Lastname, B.B. Article Title. *Journal Name* **Year**, *Volume Number*, Page Range.

ISBN 978-3-7258-6614-4 (Hbk)
ISBN 978-3-7258-6615-1 (PDF)
https://doi.org/10.3390/books978-3-7258-6615-1

Cover image courtesy of Pedro Marques de Almeida

Contents

About the Editors

Pedro Marques de Almeida

Pedro Marques de Almeida is an Associate Professor with Habilitation in the Mechanical Engineering Department of ISEL/IPL. Member of the Soft and Biofunctional Materials Group of the CENIMAT—i3N (Centre of Materials Research /Institute for Nanostructures, Nanomodelling and Nanofabrication), where he is the head of the CENIMAT NMR laboratory. He is also a member of the Unit for Innovation and Research in Engineering (UnIRE), located at ISEL, where he is President of the Scientific Council. He received a PhD and an Habilitation in Materials Science and Engineering, having the main research interest focused on Polymeric and Mesomorphic materials, Complex fluids, Solid State NMR, including MAS, Diffusion, MRI and Rheo-NMR, Liquid crystals, and liquid crystal applications, cellulose, natural materials, additive manufacturing and biomaterials. He has already supervised several Master's students, already concluded the supervision of 3 PhD students all in Materials Science and Engineering. He currently supervises one PhD student which is studying to obtain her PhD in Physics. He co-organized several international conferences mainly in the area of soft matter, liquid crystals and photonics. He has published more than 90 scientific papers, having a Scopus h-index of 25. He has participated in several national and European research projects, from which he was the leader of two. Also, he leads Work Group 2 of the COST Action CA21159 (PhoBioS), exploring light-biointerface interactions for new electronic materials and devices. He is a Member of the American Physical Society, of the International Liquid Crystal Society, and a founding member and Scientific Secretariat of the Portuguese Liquid Crystal Society.

Ana Catarina Trindade

Ana Catarina Trindade is an Assistant Professor at Atlântica – Instituto Universitário and an integrated researcher at CENIMAT—i3N. She holds a PhD in Materials Engineering, an MSc in Chemistry and a BSc in Industrial Chemistry. She conducted postdoctoral research at CEBQ/IST and at CENIMAT—i3N with competitive FCT funding, and later joined the Soft and Complex Matter Lab at NTNU (Norway) within an ERA-NET project. Her research focuses on polymeric and mesomorphic materials, cellulose nanostructures, soft-matter photonic systems, hierarchical pattern formation, and sustainable bioinspired materials. She has authored or co-authored 27 peer-reviewed articles, two book chapters and several invited presentations. Three of her works were featured on journal covers or highlighted editorially. She has participated in ten national and international R&D projects, including one as PI and one ongoing as co-PI. She has served on the organising committees of major international conferences, acts as reviewer for multiple journals and is a member of the evaluation panel of the Science Fund of the Republic of Serbia. She is a founding member of the Portuguese Liquid Crystal Society. Since 2020 she has taught more than 2500 students across Chemistry, Biochemistry, Biophysics, Materials Science and Polymer Science, including courses delivered in partnership with ITECH Lyon. She has supervised MSc theses and served on several academic juries. Her scientific and educational activities are strongly driven by sustainability, biomimetics and biopolymer nanoscience. She is also engaged in science outreach, coordinating hands-on workshops for school students. She has received national innovation awards, including the Born from Knowledge Ideas Award (ANI, 2021) and the Desafio à Inovação Award (2020).

João Canejo

João Canejo completed the PhD in Materials Engineering Ciences in 2009/02/09 by Universidade Nova de Lisboa Faculdade de Ciências e Tecnologia. He is currently teaching in the Polytechnic Institute of Portalegre and he has published 40 scientific papers in scientific journals (one distinguished as Featured Research in 2010), having a Scopus h-index of 17. He has 2 book chapters. He participated in the organizing of two international conferences. His main research interest focused on Polymeric and Mesomorphic materials, Structural Color, Production of fibers by electrospinning, Mechanical testing, and Morphlogical characterization using SEM. Participated in 1 event(s). Supervised 8 MSc dissertation(s) e co-supervised 4. He participated as Researcher in 5 project(s) and was PI of a research project.

Preface

This reprint, Polymeric Liquid Crystals and Applications, brings together a carefully selected array of the latest research on the design, characterization, and application of liquid crystal polymers (LCPs). The LCPs are unique materials that bring together the molecular self-assembly features of liquid crystals with the structural versatility of polymers. For this reason, LCPs can offer a wide range of novel properties, such as stimuli-responsiveness, tunable optical properties, and anisotropic mechanical behavior. Work in these areas ranges from the investigation of molecular relaxations, phase behavior, and synthesis strategies to applied work involving sensing, actuation, and advanced fabrication.

A definite motivation for this compilation is to provide a coherent resource that bridges theoretical insight to practical application, dealing with the growing interdisciplinary interest in smart polymeric systems. It is intended for researchers, graduate students, and engineers working in soft matter physics, polymer science, materials engineering, and sensor development. By highlighting both established knowledge and emerging directions-from auxetic elastomers and photonic sensors to confinement-directed assembly and computational modeling-this reprint hopes to further inspire innovation and collaboration in the dynamic field of polymeric liquid crystals.

Pedro Marques de Almeida, Ana Catarina Trindade, and João Canejo
Guest Editors

molecules

MDPI

Editorial

New Insights into Polymeric Liquid Crystals and Their Applications

A. C. Trindade [1,2], J. P. Canejo [2] and P. L. Almeida [2,3,*]

1 Atlântica, Instituto Universitário, Fábrica da Pólvora de Barcarena, 2730-036 Barcarena, Portugal
2 I3N-CENIMAT, Materials Science Department, School of Science and Technology, NOVA University Lisbon, 2829-516 Caparica, Portugal
3 Mechanical Engineering Department and UnIRE, Instituto Superior de Engenharia de Lisboa, Polytecnic University of Lisbon, 1959-007 Lisbon, Portugal
* Correspondence: pedro.almeida@isel.pt

Polymeric liquid crystals (PLCs) have emerged as one of the most frenetic and interdisciplinary areas of materials science, found at the crossroads of soft condensed matter, chemistry, physics, and engineering. Their ability to combine structural anisotropy with sensitivity to a broad spectrum of stimuli—such as light, temperature, and electric and magnetic fields—positions them as first-choice candidates for the development of multifunctional materials and devices [1]. The last decade has witnessed frenzied activity in the synthesis, characterization, and application of PLCs, and an indicator of this activity is the high number of articles published in high-impact journals. With this Editorial, the authors aim to provide a snapshot of recent progress in this research area, as well as the works published in this Special Issue, and look ahead to future directions based on some of the most significant recent advances in the area.

Recent developments in liquid crystal elastomers (LCEs) targeted high actuation efficiency and low operating thresholds. Lagerwall et al. introduced photoresponsive LCEs from oligomeric precursors, engineered to be active at physiological temperatures. The LCEs demonstrated efficient light-activated actuation independent of heat and, hence, are appropriate for biomedical applications such as light-actuated microfluidics and implantable devices [2]. This work follows the path of existing research regarding the regulation of phase behavior and mechanical response via molecular weight distribution and an oligomer structure. Aside from this research direction, Lin et al. developed dual-responsive LCEs through the integration of azo-functionalized polythiourethanes and polysiloxanes and mechanically interlocked molecules. The resulting systems are extremely robust and self-healing, offering a route to obtaining reconfigurable and durable soft actuators [3].

Programmable shape memory and mechanical response in polymer-dispersed liquid crystal elastomers is another innovation. Rešetič et al. introduced LC droplets in an elastomer matrix, for which there is dual-response functionality-reversible actuation in response to stimuli and shape memory in the absence of forces. Order and elasticity decoupling presents a unique pathway towards designing autonomous systems and smart soft robots [4].

PLCs are increasingly being applied in optoelectronics due to their intrinsic ability to self-assemble into photonic, anisotropic, and birefringent materials. Mysliwiec et al. described a white-light-emitting family of composites comprising PLCs and electric field-tunable photoluminescent dyes. These phase-separated structured composites can be used for future visible-light communication and lighting technologies [5]. Co-related developments include Oton et al.'s recent demonstration of directionality in laser emission via the use of blue-phase photonic crystals, based on polymer-stabilized three-dimensional LC structures [6].

In electro-optics, Singh et al. developed PDLCs containing carbon nanotube and graphene oxide additives. The resulting composites provided improved dielectric properties, higher contrast ratios, and reduced response rates (critical to the integration of smart windows and privacy glass) [7]. Similarly, Vignolini et al.'s research on switchable whiteness in LC polymer networks is the manifestation of onset dual-phase devices with transparency and reflection control through polymer confinement [8].

Azobenzene-functionalized PLCs continue to be valuable building blocks for light-switched actuation devices. Rešetič et al. investigated selectively deuterated mesogens in LCE networks for high-resolution manipulation of the interplay between the absorption of light and phase transition dynamics. This facilitated next-generation mechanical reconfiguration towards adaptive optics and soft robotics applications [4]. Meanwhile, Beeckman et al.'s research on off-axis reflective holograms through photo-alignment of chiral LCs provided a direction for high-resolution, low-power display technology [9].

Additionally, research on the 3D nanoprinting of LC-based photonic devices with large, relevant Kerr nonlinearities has been led by Wiersma et al. in the field of nonlinear optics. Such soft photonic devices, processed through two-photon polymerization, provide modulation of spatial refractive index on scales of sub-micron length—required for on-chip photonic circuitry [10]. Concurrent work focusing on modulating light–matter coupling in LCEs with embedded chromophores, as carried out by Li et al., is directed toward third-order nonlinear optical responses [11].

Charge transport and semiconducting characteristics in PLC systems have been extensively researched as well. Singh et al. researched columnar phases as active semiconducting channels in OFETs, their anisotropic transport, and morphological stability [12]. Laschat et al. also researched MR-TADF liquid crystals with narrowband emission of the mesophase itself, a new OLED technology resource [13].

Hybrid composites of PLCs and 2D materials or magnetic particles are still in vogue as a research topic. Yoon et al. researched PLC composites made of aligned MXene nanosheets with humidity-induced actuation and enhanced thermal conductivity [14]. Clark et al. focused on another related area of work, researching the magneto-optical response of ferromagnetic LCs. They found emergent behavior under rotating fields, and their research contributed towards the development of magnetically tunable devices [15]. One of the most fascinating current research areas is the synthesis of ferroelectric nematic phases and their polymer counterparts. A theory concerning PLC analogs with spontaneous polarization was developed by Mandle in a study on polar nematic order [16].

At the manufacturing level, techniques like laser patterning, field-assisted alignment, and multi-material 3D printing are enabling ultra-high-precision control of PLC domain orientation. Laser writing of nanogroove patterns for planar alignment of nematics in intricate geometries was the subject of research by Muševič et al. [17]. In parallel, in Sánchez-Somolinos et al.'s research, it was found that PLC actuators enable dynamic actuator shape reconfigurability—an approach in harmony with soft reconfigurable robotics and shape-morphing systems [18].

In our opinion, three prospective research areas are particularly relevant. The first concerns the development of green PLCs from bio-derived building blocks such as cellulose, chitin, or xanthan, a promising field in environmental sensing, biodegradable photonics, and environmentally friendly electronics. Godinho et al. have led such developments to demonstrate morphochromatic cellulose-based sensors with anisotropic optical responses [19].

Secondly, quantum photonics based on LC is an emerging field. Humar et al. have demonstrated entangled photon pairs produced from ferroelectric LC cells and proposed their application in quantum encryption and interfacing matter with light at the nanoscale [20].

Thirdly, topological engineering in PLCs is progressively becoming recognized as a viable method for the stabilization of a low-power optical memory state and metastable texture. The work carried out by Kos et al. on multistable polar textures is in line with research area, and it is also applicable to non-volatile information storage and holographic display technology [21].

The articles published in this Special Issue reflect and expand upon the current landscape of polymeric liquid crystals, addressing key scientific challenges and advancing our understanding of materials properties, novel functionalities, and potential technological applications. The study by Mattsson et al. provides a thorough investigation of the interplay between structural anisotropy and mechanical deformation, helping to elucidate the origins of auxetic behavior in LCEs. Their experimental analysis (combining dielectric spectroscopy, calorimetry, and rheology) demonstrates how strain induces configurational constraints that coincide with out-of-plane mesogen rotations, contributing to the growing body of knowledge on stimuli-responsive LCE mechanics [22]. Amaral's reflective review delivers a multidisciplinary perspective grounded in decades of work across soft matter, lyotropics, and biomolecular systems. While less focused on new data, this article plays a critical conceptual role in reminding the field of the importance of structural scaling laws, interdisciplinarity, and fundamental insights as a driver of applied research in complex soft materials [23]. Hwang et al. contribute with an original approach to chemical sensing. Their development of alcohol sensors based on cholesteric liquid crystals and carboxylate polymers enhances our understanding of photonic bandgap tuning for selectivity and sensitivity in LC-based detection systems. By demonstrating how UV curing affects performance, they address the need for simple yet effective analyte discrimination, such as discriminating between methanol and ethanol [24]. Wang et al. present an innovative bilayer PDLC structure with enhanced optical and anti-counterfeiting capabilities. Their materials engineering approach, made possible through nanoparticle doping and multilayer design, exemplifies how structural design can be leveraged to tailor electro-optical and functional responses [25]. Tokita et al. report the synthesis of side-chain liquid crystalline polyacrylates incorporating bridged stilbene mesogens, achieving nematic phases near room temperature and notable birefringence. Their polymers exhibit aggregation-induced emission and cybotactic cluster formation, offering valuable insights into π-conjugated SCLCPs and their potential in luminescent and anisotropic optical materials [26]. Kikuchi et al. present a transparent nano-PDLC system utilizing a ferroelectric Smectic A (SmAF) liquid crystal to enhance birefringence memory effects. By incorporating highly polar SmAF materials into sub-micron phase-separated structures, they achieved improved electro-optic response and retention of induced molecular orientation after field removal. This work highlights the potential of SmAF LCs in low-voltage, memory-type displays and tunable photonic devices, advancing the functionality of PDLCs beyond conventional nematic systems [27]. Choi et al. present a microlens array-imprinted polymer-dispersed liquid crystal (PDLC) film developed via a low-cost spin-coating method to enhance OLED outcoupling efficiency. Their flexible, high-haze PDLC structure achieves a 37.5% improvement in light extraction without compromising electrical performance. This work demonstrates the practical integration of soft lithography and LC-based scattering layers for flexible optoelectronics, pointing toward scalable roll-to-roll manufacturing of next-generation OLED displays [28]. Wie et al. report a hybrid top-down/bottom-up strategy for fabricating highly aligned liquid crystalline polymeric fibers using spatial confinement. By thermally oligomerizing LC monomers within microchannels and subsequently photopolymerizing them, the authors demonstrate the controlled transition from spherical micelles to linear, defect-free nematic fibers. This confinement-driven self-assembly offers an alternative to high-shear or electrospinning techniques and paves new pathways for the scalable production of anisotropic LC fibers for soft actuators and responsive materials [29]. Finally, Zakaria et al. present a detailed structure–property investigation of Schiff-base ester liquid crystals featuring pyridyl and phenyl heterocycles. Through experimental mesophase characterization and complementary DFT analysis, the authors demonstrate how mesogenic core orientation and terminal alkyl chain length critically influence mesophase type and thermal stability. The study identifies dipole moment, planarity, and π–π stacking potential as key predictors of nematic versus smectic behavior, offering valuable design rules for tailoring new mesogenic architectures with tunable electro-optic properties [30].

Collectively, these papers present a coherent picture of the developments in polymeric liquid crystal research, from providing a fundamental understanding of mechanical relaxations and auxetic behavior in elastomers to innovative applications in sensing, photonics, soft actuation, and smart interfaces. They also demonstrate the importance of coupling molecular design with advanced processing techniques (such as electrospinning, surface grafting, or microfluidic templating) to enable next-generation materials. Future work in this area will undoubtedly build on the insights offered here, exploring hybrid architectures, multiresponsive functionalities, and scalable manufacturing methods. Polymeric liquid crystal science is growing and developing. By integrating dynamic molecular self-assembly with mechanical response and optical function, PLCs offer an unprecedented versatility of material platform. The research documented in this Special Issue not only heralds recent achievements but also predicts the exciting future prospects of PLCs as enabling technologies for future generations of smart, responsive, and multifunctional systems.

Author Contributions: A.C.T., J.P.C. and P.L.A.: methodology, validation, formal analysis, investigation, data curation, writing—original draft preparation, and writing—review and editing. All authors have read and agreed to the published version of the manuscript.

Funding: Funding was provided by FCT (Fundação para a Ciência e a Tecnologia, I.P.) in the scope of the projects LA/P/0037/2020, UIDP/50025/2020, and UIDB/50025/2020 of the Associate Laboratory Institute of Nanostructures, Nanomodelling and Nanofabricationi3N.

Conflicts of Interest: The authors declare no conflicts of interest.

References

1. Almeida, A.P.C.; Canejo, J.P.; Fernandes, S.N.; Echeverria, C.; Almeida, P.L.; Godinho, M.H. Cellulose-Based Biomimetics and Their Applications. *Adv. Mater.* **2018**, *30*, 1703655. [CrossRef]
2. Najiya, N.; Popov, N.; Jampani, V.S.R.; Lagerwall, J.P.F. Continuous Flow Microfluidic Production of Arbitrarily Long Tubular Liquid Crystal Elastomer Peristaltic Pump Actuators. *Small* **2023**, *19*, 2204693. [CrossRef] [PubMed]
3. Trung, N.T.; Chiu, C.; Cuc, T.T.K.; Khang, T.M.; Jalife, S.; Nhien, P.Q.; Hue, B.T.B.; Wu, J.I.; Li, Y.; Lin, H.-C. Tunable Nano-Bending Structures of Loosened/Tightened Lassos with Bi-Stable Vibration-Induced Emissions for Multi-Manipulations of White-Light Emissions and Sensor Applications. *Adv. Mater.* **2024**, *36*, 2311789. [CrossRef] [PubMed]
4. Bobnar, M.; Derets, N.; Umerova, S.; Domenici, V.; Novak, N.; Lavrič, M.; Cordoyiannis, G.; Zalar, B.; Rešetič, A. Polymer-Dispersed Liquid Crystal Elastomers as Moldable Shape-Programmable Material. *Nat. Commun.* **2023**, *14*, 764. [CrossRef]
5. Szukalska, A.; Szukalski, A.; Adaszynski, M.; Mysliwiec, J. White Lasing and White Fluorescence from the Simplified Two-Dyes Organic System. *Adv. Opt. Mater.* **2023**, *11*, 2300266. [CrossRef]
6. Oton, E.; Cigl, M.; Morawiak, P.; Mironov, S.; Bubnov, A.; Piecek, W. All-Optical 3D Blue Phase Photonic Crystal Switch with Photosensitive Dopants. *Sci. Rep.* **2024**, *14*, 9910. [CrossRef]
7. Singh, B.P.; Hwang, S.-J. Unveiling the Potential of Polymer Cholesteric Liquid Crystal Interpenetrating Networks as a Label-Free Alcohol Biochemical Sensor. *Analyst* **2024**, *149*, 3456–3467. [CrossRef]

8. Jacucci, G.; Schertel, L.; Zhang, Y.; Yang, H.; Vignolini, S. Light Management with Natural Materials: From Whiteness to Transparency. *Adv. Mater.* **2021**, *33*, 2300266. [CrossRef]
9. Ropač, P.; Hsiao, Y.; Berteloot, B.; Ravnik, M.; Beeckman, J. Material-Constrained Optimization of Liquid Crystal-Based Holograms. *Adv. Opt. Mater.* **2024**, *12*, 2400972. [CrossRef]
10. Zanotto, S.; Sgrignuoli, F.; Nocentini, S.; Martella, D.; Parmeggiani, C.; Wiersma, D.S. Multichannel Remote Polarization Control Enabled by Nanostructured Liquid Crystalline Networks. *Appl. Phys. Lett.* **2019**, *114*, 201103. [CrossRef]
11. Zhou, L.; Chen, H.; Li, K. Optically-Responsive Liquid Crystal Elastomer Thin Film Motors in Linear/Nonlinear Optical Fields. *Thin-Walled Struct.* **2024**, *202*, 112082. [CrossRef]
12. Shah, A.; Vishwakarma, V.K.; Lhouvum, N.; Sudhakar, A.A.; Kumar, P.; Srivastava, A.K.; Dubois, F.; Chomchok, T.; Chattham, N.; Singh, D.P. Pyrazino [2,3-g]Quinoxaline Core-Based Organic Liquid Crystalline Semiconductor: Proficient Hole Transporting Material for Optoelectronic Devices. *J. Mol. Liq.* **2024**, *393*, 123535. [CrossRef]
13. Knöller, J.A.; Müller, F.; Matulaitis, T.; dos Santos, J.M.; Gupta, A.K.; Zysman-Colman, E.; Laschat, S. MR-TADF Liquid Crystals: Towards Self Assembling Host–Guest Mixtures Showing Narrowband Emission from the Mesophase. *Chem. Sci.* **2024**, *15*, 18022–18030. [CrossRef]
14. Lee, C.; Park, S.M.; Kim, S.; Choi, Y.-S.; Park, G.; Kang, Y.C.; Koo, C.M.; Kim, S.J.; Yoon, D.K. Field-Induced Orientational Switching Produces Vertically Aligned Ti3C2Tx MXene Nanosheets. *Nat. Commun.* **2022**, *13*, 5615. [CrossRef] [PubMed]
15. Shuai, M.; Klittnick, A.; Shen, Y.; Smith, G.P.; Tuchband, M.R.; Zhu, C.; Petschek, R.G.; Mertelj, A.; Lisjak, D.; Čopič, M.; et al. Spontaneous Liquid Crystal and Ferromagnetic Ordering of Colloidal Magnetic Nanoplates. *Nat. Commun.* **2016**, *7*, 10394. [CrossRef] [PubMed]
16. Gibb, C.J.; Hobbs, J.; Nikolova, D.I.; Raistrick, T.; Berrow, S.R.; Mertelj, A.; Osterman, N.; Sebastián, N.; Gleeson, H.F.; Mandle, R.J. Spontaneous Symmetry Breaking in Polar Fluids. *Nat. Commun.* **2024**, *15*, 5845. [CrossRef] [PubMed]
17. Jagodič, U.; Vellaichamy, M.; Škarabot, M.; Muševič, I. Surface Alignment of Nematic Liquid Crystals by Direct Laser Writing of Photopolymer Alignment Layers. *Liq. Cryst.* **2023**, *50*, 1999–2009. [CrossRef]
18. López-Valdeolivas, M.; Liu, D.; Broer, D.J.; Sánchez-Somolinos, C. 4D Printed Actuators with Soft-Robotic Functions. *Macromol. Rapid Commun.* **2018**, *39*, 1700710. [CrossRef]
19. Fernandes, S.N.; Almeida, P.L.; Monge, N.; Aguirre, L.E.; Reis, D.; de Oliveira, C.L.P.; Neto, A.M.F.; Pieranski, P.; Godinho, M.H. Mind the Microgap in Iridescent Cellulose Nanocrystal Films. *Adv. Mater.* **2017**, *29*, 1603560. [CrossRef]
20. Sultanov, V.; Kavčič, A.; Kokkinakis, E.; Sebastián, N.; Chekhova, M.V.; Humar, M. Tunable Entangled Photon-Pair Generation in a Liquid Crystal. *Nature* **2024**, *631*, 294–299. [CrossRef] [PubMed]
21. Čopar, S.; Kos, Ž. Many-Defect Solutions in Planar Nematics: Interactions, Spiral Textures and Boundary Conditions. *Soft Matter* **2024**, *20*, 6894–6906. [CrossRef]
22. Raistrick, T.; Reynolds, M.; Gleeson, H.F.; Mattsson, J. Influence of Liquid Crystallinity and Mechanical Deformation on the Molecular Relaxations of an Auxetic Liquid Crystal Elastomer. *Molecules* **2021**, *26*, 7313. [CrossRef] [PubMed]
23. Amaral, L.Q.d. Supramolecular Aggregates: Hardness Plus Softness. *Molecules* **2021**, *26*, 4233. [CrossRef] [PubMed]
24. Yeh, T.-Y.; Liu, M.-F.; Lin, R.-D.; Hwang, S.-J. Alcohol Selective Optical Sensor Based on Porous Cholesteric Liquid Crystal Polymer Networks. *Molecules* **2022**, *27*, 773. [CrossRef]
25. Wu, Y.; Zhao, Y.; Li, X.; Gao, H.; Guo, Z.; Wang, D.; Luan, Y.; Wang, L. Preparation and Characterization of Bilayer Polymer-Dispersed Liquid Crystals Doped with Gd_2O_3 Nanoparticles and Rhodamine B Base Fluorescent Dye. *Molecules* **2024**, *29*, 1126. [CrossRef]
26. Konishi, G.; Sawatari, Y.; Iwai, R.; Tanaka, T.; Shimomura, Y.; Tokita, M. Synthesis of Side-Chain Liquid Crystalline Polyacrylates with Bridged Stilbene Mesogens. *Molecules* **2024**, *29*, 5220. [CrossRef]
27. Yamaguchi, M.; Matsukizono, H.; Okumura, Y.; Kikuchi, H. Nanostructured Polymer-Dispersed Liquid Crystals Using a Ferroelectric Smectic A Liquid Crystal. *Molecules* **2024**, *29*, 4837. [CrossRef]
28. Lim, S.; Ahn, H.-S.; Jang, E.-J.; Boo, S.-Y.; Gasonoo, A.; Gwag, J.-S.; Lee, J.-H.; Choi, Y. Polymer Dispersed Liquid Crystal Imprinted by Microlens Array for Enhanced Outcoupling Efficiency of Organic Light Emitting Diode. *Molecules* **2023**, *29*, 73. [CrossRef]
29. Kim, J.G.; Lee, J.G.; Wie, J.J. Confinement-Induced Fabrication of Liquid Crystalline Polymeric Fibers. *Molecules* **2022**, *27*, 5639. [CrossRef]
30. Nada, S.; Hagar, M.; Farahat, O.; Hasanein, A.A.; Emwas, A.-H.; Sharfalddin, A.A.; Jaremko, M.; Zakaria, M.A. Three Rings Schiff Base Ester Liquid Crystals: Experimental and Computational Approaches of Mesogenic Core Orientation Effect, Heterocycle Impact. *Molecules* **2022**, *27*, 2304. [CrossRef]

molecules

MDPI

Review

Supramolecular Aggregates: Hardness Plus Softness

Lia Queiroz do Amaral

Instituto de Física, Universidade de São Paulo, São Paulo CEP 05508-090, Brazil; amaral@if.usp.br

Abstract: The properties of supramolecular aggregates cross several disciplines, embracing the sciences of nature and joining theory, experiment, and application. There are few articles centering on the problems of interdisciplinarity, and this paper gives an alternative approach, starting with scientific divulgation, bringing concepts from their origin, to facilitate the access of young scientists to the scientific content. Didactic examples are taken from the experience of the author in changing directions of research due to several circumstances of life (including maternity), starting from the view of a rigorous student of physics and evolving to several subjects in chemistry. The scientific part starts with concepts related to nuclear interactions, using the technique of neutron scattering in reactors, and evolves to research in molecular physics. Finally, it arrives at the academic context, with research in condensed matter physics, with X-ray and other techniques, starting with detergents forming nematic lyotropic liquid crystals and the phase transition sequence of isotropic to nematics to hexagonal. The scientific subjects evolved to biological and bio-inspired liquid crystals, including DNA and also specific lipids and phospholipids in biomimetic membranes. Special attention is given to the question of distribution of matter in these complex systems and the non-trivial connections between biochemistry, structures, auto-aggregation, and biology.

Keywords: interdisciplinary perspective; basic concepts; molecular physics; lyotropic liquid crystals; molecular biophysics; biology

1. Introduction

Hardness and softness are properties of materials on the macroscopic scale but are the result of interactions at the molecular level. Solids are the result of the existence of an atomic structure with atoms fixed at defined positions, while softness results from the fact that atoms have freedom of motion. The degree of freedom and the existence of spatial correlations between different atoms define the differences in macroscopic behavior. For instance, a diamond crystal and the graphite of a pencil are both made of carbon atoms, which exist also as a gas. Water is a liquid existing in the three states of matter by temperature variation, with very complex interactions between only two types of atoms, hydrogen and oxygen. Basic research in nature sciences aims to understand the interactions in matter, while engineering focuses on controlling processes for uses in human activities.

Following the transformations due to the industrial revolution, in the second half of the 18th century, the study of solid materials was introduced formally, by the middle of the 19th century, in engineering courses in Germany and England, and also in the USA. The needs of the nascent industry required knowledge of elasticity and material resistance, together with metallurgy and crystallography. Focus was also given to changes in the material properties with changes in the macroscopic variables of a system, for example, changes from a solid to a liquid, from a liquid to a gas, or between different phases of metals and alloys. Basic books on these fundamental issues are necessary for graduate students to enter these fields, especially from a theoretical point of view [1,2]. Note that [1] has the view of metallurgy (not the exactness of crystallography) while [2] is for a general audience, and none of them are "for crystallographers", and that is a first challenge towards multidisciplinary issues.

Scientific studies of supramolecular aggregates evolved initially in two different directions: colloid science and supramolecular chemistry, both starting before the turn of the 20th century. In the last decades, this interdisciplinary field, with the study of intermolecular bonding, has included systems that self-assemble, at the triple meeting point of chemistry, biology, and physics [3], and has reached the 21st century with an incredible power in terms of both basic knowledge and technological applications. It is indeed a long history, connected to deep transformations in human societies!

There are many specialized articles on specific aspects of subjects in the frontiers of science and their technological applications. However, there are few articles centering on the problems of interdisciplinarity, with a comprehensive approach for students and beginners. Knowledge evolves in each discipline starting from its specific criteria, and there are no conceptual bridges among them, so conditions for synthesis are not trivial.

The approach between neighboring research fields is in general obtained from methodologies and techniques which can be used in both fields, but this, in general, means that close disciplines may diverge, even when using the same techniques. The best example is the parallel existence of physical chemistry and chemical physics, with an uneasy dialogue between them. The problem with biology is much bigger, since there is not even a basic conceptual starting point in common.

This paper aims to offer an alternative approach to this difficult problem, focusing on basic concepts necessary to enter interdisciplinary fields, with an initial approach of scientific divulgation.

Studies of colloids started in the early 1800s. The word is derived from the Greek "kola" (glue) and was used to distinguish colloidal solutions from solutions with crystallized particles (such as sugar and salt). A colloidal solution is a heterogeneous system consisting of a mechanical mixture of particles with sizes between 1 and 1000 nm, in a continuum medium (solid, liquid, or gas), with behavior that is intermediate between a true solution and particles dispersed in a medium (suspension, emulsion, or foam). This interdisciplinary field evolved to studies of the reactivity of surfaces and interfaces, making a bridge between the macroscopic level and the atomic and molecular levels. It can be said that colloid and interface science dealt with nanoscale objects for nearly a century before the term *nanotechnology* was coined [4]. Interface and colloid science have applications and ramifications in the chemical industry, pharmaceuticals, biotechnology, ceramics, minerals, nanotechnology, and so on [5].

Beginning in the 19th century, liquid crystals (LC) also have a long history. In 2013, their 125th anniversary was celebrated [6]. This article of fewer than five pages gives 10 basic references to those wishing to enter the field, providing the foundations on thermotropics (temperature-dependent), lyotropics (composition-dependent), and applications (from displays to biology).

Supramolecular chemistry can be considered to begin in the late 1800s when the German organic chemist Hermann Emil Fisher, studying purines and sugar, suggested that enzyme/substrate interactions take the form of a "lock and key". He received the Nobel Prize in 1902. The fundamental principles of molecular recognition and host–guest chemistry developed, and in the early 20th century, intermolecular bonds started to be understood in gradually more detail, through a merging process with quantum physics, a difficult process [7].

In the early 1900s, physics found its way focusing on quantum mechanics and turning to the microscopic world below the level of the hydrogen atom, while chemistry focused on molecules and above, turning towards the human world, and chemical industries helped to change human societies. Quantum chemistry dominated the scene in chemistry until the discovery of the structure of DNA in the 1950s. The second half of the 1900s saw the emergence of biochemistry and molecular biology—a serious attempt at integration of the natural sciences, posing an enormous challenge for graduate education [8].

Let us now turn back to some specific supramolecular aggregates: polymers and macromolecules, which are not the same thing. The word "polymer" (from the Greek polys meaning "many" and meros meaning "part") was first introduced in 1833 by the Swedish chemist Jöns Jakob Berzelius, while the German chemist Hermann Staudinger felt it necessary to coin the word "macromolecule" in 1922 to describe large covalently bonded organic chain molecules containing more than 103 atoms [9].

Since pre-history, human activities have involved the use of materials, both natural and modified, many of them containing polymers (such as wool, cotton, fibers, etc.). Chemical manipulation of such raw materials started in the 19th century, without knowledge of their nature. Initial rationalization was proposed by the Scottish chemist Thomas Graham (Royal Society medals in 1838 and 1850), considering them as colloidal aggregates of small molecules, held together by unknown forces. Even without theories, the potential of polymers in industry was grasped and used. The German chemist Hermann Staudinger proposed in 1920 that polymers were in fact long chains of atoms linked by covalent bonds, receiving the Nobel Prize in 1953. The subsequent discovery of biopolymers, natural polymers produced by the cells of living organisms, opened the Pandora's box of biochemistry and molecular biology.

Main-chain and side-chain polymers can be both lyotropic and thermotropic LC polymers. The chemical structure of a polymer backbone, either as a single polymer or in bulk, is often not suitable to be used in practical applications. Polymer–dye conjugations are one of the common examples of polymer modifications. They represent a crucial step for imaging in optical microscopy or for tracing and marking macromolecules. A comparison of recent original strategies to conjugate polymers to ligands and dyes (e.g., for detection or targeting purposes in health care applications) can be found in [10].

Up to here, this survey has given just a snapshot of the problems faced nowadays by young scientists at the crossing of the so-called sciences of nature. This review turns now to my own trajectory, as an example of the difficulties faced in the transition from the student level to the professional level in science, particularly for a woman, in a peripheral country, entering middle school around 1955 and good in mathematics.

My trajectory touched eventually on polymers, but that was not my expertise. I have worked mainly at the interfaces of physics/chemistry/biology/education, and at the interface of natural sciences/human issues. I do not mention all the research developed along my career, but several of my own papers are given as references, when having direct relation to the subjects discussed.

As a student of physics, I started a successful initiation of research in nuclear physics, and immediately after graduation I secured a job working in a research reactor, related to the technique of slow neutron scattering, where I stayed for 12 years. However, I quit that job after my daughter was born. Later on, it was possible for me to restart from zero, with a provisional part-time job at the Institute of Physics of USP (IFUSP), working in other directions. Only several years later was I able to return to the data from the "neutron time", when I was again in full-time research in condensed matter physics, with tenure at IFUSP.

This review emphasizes interdisciplinary concepts from the beginning of my career, which are good for didactical explanations of the changes of direction I needed to face. This is included in Section 2, along with some more personal issues, while Section 3 deals with more objective scientific discussions on obtained results from the neutron time. Section 4 deals with the project I started in IFUSP, creating a new X-ray Laboratory of Crystallography and initiating research on lyotropic liquid crystals, while Section 5 takes the discussion in the direction of biology, focusing more on the present.

2. Theory vs. Experiment

The discussion of the relative importance of theory and experiment has no trivial answer; knowledge is constructed from both. Sometimes an experiment is planned to test a theory, but many times an experimental result is obtained for practical reasons, and the search for a theoretical explanation comes later on. This depends also on the field being considered, and on its historical development.

It should be stressed that what are called "exact sciences" are the result of the union of physics with chemistry at the turn of the 20th century. Modern physics came after the definition of chemistry as a true science, with the construction of the periodic table of elements. Physical theories are more focused on mathematical rigor and "elegance", while chemical theories are usually more connected to experimental results.

Around the year 1940, areas of interest to physics and engineering were grouped together as solid state physics, with the development of semiconductors leading to micro-electronics. Materials science started to become an extremely vast multidisciplinary field, defined by the triangle structure–processing–properties, with applications in engineering and industry.

Condensed matter physics started around 1960, when the study of liquids started to become relevant. Simple liquids, made of a single component, started to be investigated by physics, but solutions with multiple components were subjects of physical chemistry.

This review will now consider the perspective of a young female (YF) student of physics, very rigorous in mathematics, initiating scientific research in 1960, in the middle of ideological conflicts with external reality. This section will chronologically follow the YF's experiences (***important points stressed in bold italics***).

2.1. In Nuclear Physics

YF started to do research in a large project on monazite sands from Brazil, connected to the local chemistry industry, with the specific aim to study, by slow neutron transmission, the total cross-sections of some rare earths of interest to nuclear energy in connection with the production of thorium [11]. The experimental steps in the nuclear research reactor included getting the samples; conditioning them in sample holders; making all measurements at the reactor; understanding how the neutron beam was obtained and how to change its energy; using a crystal monochromator, and a neutron mechanical velocity selector, both locally constructed; making transmission measurements with the detector; and using the available electronics. Of particular interest was the resonance curve of lutetium (Lu) as function of neutron energy. In parallel, YF attended classes on nuclear physics and quantum mechanics (QM), and also the first course on computers given by IBM in Brazil, learning to program in Fortran and using an IBM 1620 computer for data analysis.

Then, however, YF realized that the theoretical courses did not directly help the analyses, as there was no clear route to join theory with obtained data. It was clear that Breit–Wigner theory [12] should be used to interpret the experimental results for the resonance in Lu. However, QM scattering theory had possible solutions only in the first Born approximation, which was not valid for slow neutrons. The problem seemed untreatable since the true nuclear potential was also unknown. YF found out that a genial solution is the pseudo-potential introduced by Fermi [13], which solves the untreatable problem of scattering of slow neutrons by using a delta function and the known experimental result for the scattering length. With such a strategy, the project arrived at a result useful to nuclear energy, presented by YF at an international local congress promoted by the Brazilian Atomic Energy Committee together with the International Atomic Energy Agency (IAEA, 1963). Several years later, it resulted in an academic publication [14].

The solution is to escape from an untreatable, exact theory by using a correct conceptual solution together with an experimental result.

2.2. In Molecular Physics

YF, after research on nuclear physics with slow neutrons in Brazil, was sent on a fellowship with IAEA to Sweden, alone in 1964, to work in the small group of Prof. Karl-Erik Larsson at the Royal Institute of Technology, Department of Reactor Physics, in Stockholm, and stayed there for 15 months. The samples were hydrogenous liquids, studied by cold neutron incoherent scattering, and the experimental setup included a filter of beryllium refrigerated by liquid nitrogen inside the reactor beam, used to extract the cold neutrons, while a chopper with time-of-flight made the energy analysis. Not in a transmission geometry but with angular variation in order to also measure momentum transfer.

YF's personal life was very difficult in Sweden, an experience of hard loneliness, but also of freedom and independence.

After some time reading the literature, YF could grasp the theoretical problem. It was the QM theory for correlations in space and time and Born approximation scattering in systems of interacting particles, the Van Hove formalism [15], using the Fermi pseudo-potential for the neutron–nucleus interaction. However, this was extended to all neutron energies by describing the Born approximation scattering in terms of the time-dependent pair-distribution function G (r,t), a very natural extension of the conventional static g(r) function.

Exact, beautiful, and understandable, but to apply that theory to complex hydrogenous molecular systems was again an intractable problem!

Prof. Larsson was working on a theoretical model for the derivation of a neutron-scattering cross-section for quasi-elastic scattering from a complex hydrogenous liquid, separating the proton motion within the molecule from the motion of protons between molecules. The foreign students, however, worked on data from specific samples. The Be filter produced an incident beam with a sharp edge, due to a Brag peak [16] and a continuum cold spectrum for higher wavelengths. Analyzed samples showed a broadening of the sharp edge, related to a quasi-elastic scattering. Hydrogenous liquids also presented an inelastic part, which was not focused by the group.

The work in Sweden was published later on in Physical Review. It included a first section by Larsson with Bergsted, on theory [17], and a second section by Larsson and the foreign students, on experimental results related to pentane and n-propyl alcohol [18].

Only after seeing the final paper, one year later, could YF really understand what was done by the whole group, and her role in all of it.

2.3. In Getting Academic Degrees

Back home, YF focused on transforming the previous experimental results on rare earths into an academic paper [14].

At the same time, results started to be obtained with the new equipment, which was similar to the Swedish equipment, focusing first on the calibration and resolution of the chopper time-of-flight spectrometer, with a detailed mathematical analysis of the problem. These results were published as a paper [19]. Neutron data was also obtained in solid materials (uranium oxide and polycrystalline iron) and in a large series of hydrogenous liquids. Several papers of the reactor group were presented in the first meeting of the Brazilian Physical Society (1966).

In 1968, a big change in internal politics occurred in academic life in Brazil. The European system, based on academic chairs with professors and their assistants, was being replaced by the American system, with post-graduate courses, requiring qualifying examinations and credits before Master and Doctorate degrees. At the same time, all the departments of physics existing in faculties had to move towards independent institutes of physics.

YF took the opportunity to get a master's degree in Nuclear Science and Engineering, wrote a dissertation based on the published paper [19], presented and defended it in the new American system (June 1969), and was taken on to complete her Ph.D. in Physics, still in the European system, with a rigid deadline: the end of 1972.

For the thesis, YF found a suitable compound in a handbook of chemical products: a plastic crystal with phase transitions near room temperature. Data was obtained through two years of experimental work. Already pregnant, YF worked hard to write the thesis, and gave it the flavor of molecular physics, also analyzing inelastic neutron differential scattering.

The Ph.D. thesis was defended at the end of November 1972, and maternity occurred five days later. Since it was impossible to reconcile her job hours with her personal life, YF decided to quit the nuclear reactor to stay at home.

2.4. In Changing Directions

The change in direction meant a definite change in technique since slow neutrons were available only at the reactor. The opportunity to start a part-time job at IFUSP meant giving classes on basic physics (theory and laboratories) for the initial two years of the engineering courses. This also entailed a proposal for research, starting a new X-ray crystallography laboratory (CrysLab) together with a colleague that had just finished a Ph.D. focused on defects in crystals.

The basic theory for the interaction with matter is rather different. Slow neutrons interact with the nuclei treated with the Fermi pseudo-potential, while X-rays interact with the electrons via classical electromagnetism. Furthermore, slow neutrons interact mostly with hydrogen, via incoherent scattering, while X-rays interact with crystals via coherent scattering. The previous expertise of YF with neutrons was of little help.

For the Ph.D. thesis, YF had some interaction with the chemistry department of USP, since in physics, the focus was on the theoretical QM of the H atom, not of molecules. Looking for a new subject of research, YF again went there, and, by chance, attended the seminar of a Canadian chemist who was visiting the NMR group and who worked with lyotropic nematic liquid crystals.

From this beginning, added to the previous experience with hydrogenous liquids and plastic crystals, YF was able to define a research project on lyotropic liquid crystals (in close connection with bio-membranes). It would be a collaboration with the NMR group of the Chemistry Institute, where the samples (aqueous mixtures of water/detergent/additives) were prepared. The NMR research was focused on the structure of molecules (such as benzene) oriented by the lyotropic systems, while the IFUSP project proposed X-ray structural study of the lyotropic phases.

The whole project had no connection with the research done at IFUSP. It was presented to a financial agency within a general proposal to open IFUSP to new areas of research. It took two years for the equipment to arrive, and during this time, YF returned to the neutron data, in order to develop scientific papers. The result is objectively presented in the next section of this review.

3. Conformations of Hydrocarbon Chains by Slow Neutron Scattering

This subject was chosen to receive special attention in this review since it provides an opportunity to discuss objective results obtained with the technique of neutron scattering in the years' 1960s.

Studies of rotational and vibration degrees of freedom by molecules using spectroscopic methods were already well established by the year 1960 [20]. However, for those working with hydrogenous materials with slow neutrons using research reactors, an alternative approach became available through analysis of the total neutron scattering cross section per H atoms, σ_S/H, with a defined slope as a function of neutron energy [21]. The model, based on the Fermi pseudo-potential, describes the scattering in terms of two molecular parameters: an effective proton mass for translation and rotation, and a vibration constant equal to the mean square zero-point vibration displacement of the proton, with good results in the cases of CH_4 and H_2. Measurements of ammonium halides have also shown a definite correlation between the slope of the cross-section of low-energy neutrons and the rotational freedom of the ammonium ions NH_4 [22].

3.1. In Hydrogenous Materials in São Paulo

The neutron transmission methodology has been used in the study of rotational freedom of a large series of hydrogenous liquids in the research reactor in São Paulo [23].

The total cross-section $\sigma = ln\ (T^{-1})/n$, where T is the measured neutron transmission and n is the number of molecules/cm^2, is mostly incoherent. The scattering cross-section per H atom σ_s/H is obtained by correcting for absorption and dividing by the number of H atoms in the molecule. The neutron transmission methodology is based on a calibration curve correlating the slope of σ_s/H to the activation energy of the rotation being considered for a series of molecules, using the Krieger–Nelkin (KN) model [21], eventually with improvements. The slopes of σ_s/H, in fact, are not directly connected to the barriers for rotation, but rather related to the energy states below the barrier available for energy-gain scattering by the cold neutrons.

The alcohol molecules in the liquid phase are connected via hydrogen bonding, and differences in the measured slopes reveal free rotation and hindered rotations in CH_3 and CH_2 groups [24]. In the cases of methanol, at room temperature, the neutron quasi-elastic scattering results give relaxation times, and the activation energy is in agreement with results from microwave methods [25].

3.2. *In the Polymer PDMS*

With the simple neutron transmission methodology, a study of the CH_3 rotations in the polymer polydimethylsiloxane (PDMS) at room temperature has been also performed at the São Paulo reactor, and is discussed here in some detail. This silicone, $(C_2H_6OSi)_n$, has linear units $(CH_3)_2SiO$ and terminal groups $(CH_3)_3SiO$. The almost free rotation of the CH_3 groups was already accepted, but evidence was found also for chain rotation and self-diffusion. NMR results could not exclude hindered rotations of CH_3 with a small barrier. Slow neutron transmission, a very simple measurement, could, however, give some answers to the problem. This work aimed to clarify different interpretations of the inelastic neutron spectra. In the published paper [26] a figure shows the curve relating the slopes to the KN mass parameter. Here the measured σ_s/H for PDMS as a function of the neutron wavelength λ is given in Figure 1, showing a very good linearity over the range 5–10 Å, with a slope 12.2 ± 0.2 barns/Å.

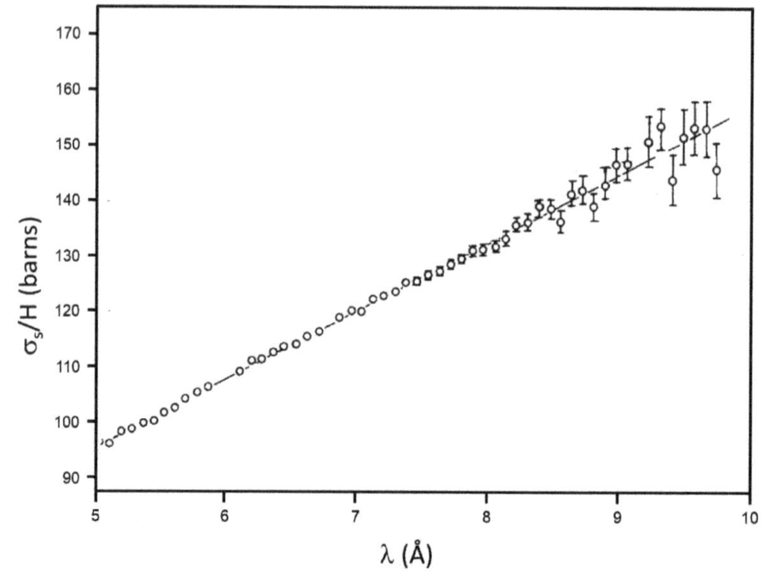

Figure 1. Total scattering cross-section per H atom, σ_s/H, at room temperature, for PDMS (polymer polydimethylsiloxane) as a function of the neutron wavelength λ. Experimental points and adjusted straight line, with slope related to the barrier for CH_3 rotations.

Known calibration curves [22,23,26] give a limit of 0.4 kcal/mole for the barrier hindering methyl rotation in PDMS, since corrections for other low-frequency motions of the molecules in the liquid state would result in a small reduction in the slopes attributed to internal rotation, but these corrections are difficult to estimate. The paper [26] includes some considerations on other contributions, due to larger rotational mass. The conclusion is indication of practically free rotation of CH_3 groups about their C_3 symmetry axes. Such unusual freedom was attributed to the greater radius of the silicon atom compared to the C atom, so that the Si-C bond is longer than the C-C bond, and also to the separation between methyl groups, due to the presence of oxygen atoms along the chain skeleton.

It is interesting to mention that this simple transmission result was cited afterwards in a paper comparing the techniques of neutron inelastic scattering and NMR for PDMS below room temperature [27]. Satisfactory agreement of the two experimental techniques is achieved as regards the activation energy and the pre-exponential factor of the Arrhenius approach for the correlation time and jump time. It can be added that PDMS is nowadays used for such large and numerous applications that it is no longer restricted to academic research, and has become the target of industrial design [28].

3.3. *In a Plastic Crystal*

It is convenient to turn now to the study of a molecule classified as a "globular compound", with a plastic phase intermediary between solid and liquid, where rotations of CH_3 are expected to be free: tert-butanol (C_4H_9OH) or $(CH_3)_3$-C-OH. The study of its molecular dynamics by slow neutron scattering was the object of a Ph.D. thesis in the research reactor in São Paulo [29], with two articles published afterwards [30,31]. A combination of measurements of the total cross-section, by transmission, and of the double differential scattering cross-section as a function of changes in both momentum and energy transfer, was performed with temperature variation. Neutron transmission results were obtained in the temperature interval 0–40 °C, where the phase transitions occur [30], and later, the quasi-elastic and the inelastic scattering provided more information on the molecular dynamics [31].

Abrupt changes in σ_S/H as a function of temperature may be attributed to the inelastic component and give information on changes in the freedom of motion of molecules and molecular groups at state and phase transitions. Slow neutrons have a short interaction time (ca. 10^{-12} s) and are especially valuable for studying CH_3 rotations because of the high hydrogen scattering cross-section and large amplitude of motion of the protons. In the case of hindered rotations, the CH_3 fundamental torsional frequency is easily seen in inelastic neutron scattering spectroscopy, while it is very difficult to detect in infrared and Raman spectra. In view of such characteristics, the molecular dynamics of tert-butanol in two crystalline phases (with a transition at 13 °C), and in the liquid state (melting transition at 23 °C), were investigated by cold neutron scattering.

Here, the transmission results are discussed in some more detail. The article [30] shows only the transmission measurements versus temperature as averages over 22 series of cooling and/or heating. Results indicate an abrupt step in cooling and a softer one in heating at the change of state, as well as indication of a third crystalline form. Here, the focus is on the barrier to the rotation of the CH_3 groups, which requires consideration of the bulk density of the sample, with discontinuities at the state and phase transitions, as shown in Table I in [30]. Since it was not possible to know if differences in the sample thickness occurred at the state transition, the two limiting values were considered in the solid state.

Figure 2 shows the results for σ_S/H as a function of neutron wavelength λ, for the liquid state and for the two limiting values of the solid state. Straight line adjustment is well defined over the whole λ interval in the liquid state, giving a slope of 8.6 ± 0.2 barns/Å. In the solid state, the two limiting values are very near, the average value was considered, and the adjustment for $\lambda > 5$ Å gives a slope of 5.7 ± 0.3 barns/Å. The passage from the slope to the barrier V was made according to two different calibration curves: one due to rotation of NH_4 in the solid state [22], and the other due to rotation of CH_3 in associated liquids [23]. The different values correspond to the existence of whole molecule movements in the liquid phase, and the conclusion is that, for tert-butanol, the calibration curve given in [22] should be used for the solid phase, giving V = 3.8 ± 0.5 kcal/mol, while the calibration curve given in [23] should be used in the liquid phase. The overall conclusion from the transmission measurements is that the internal rotation of the methyl group is not sensible for the state and phase transitions, being practically independent from intermolecular forces.

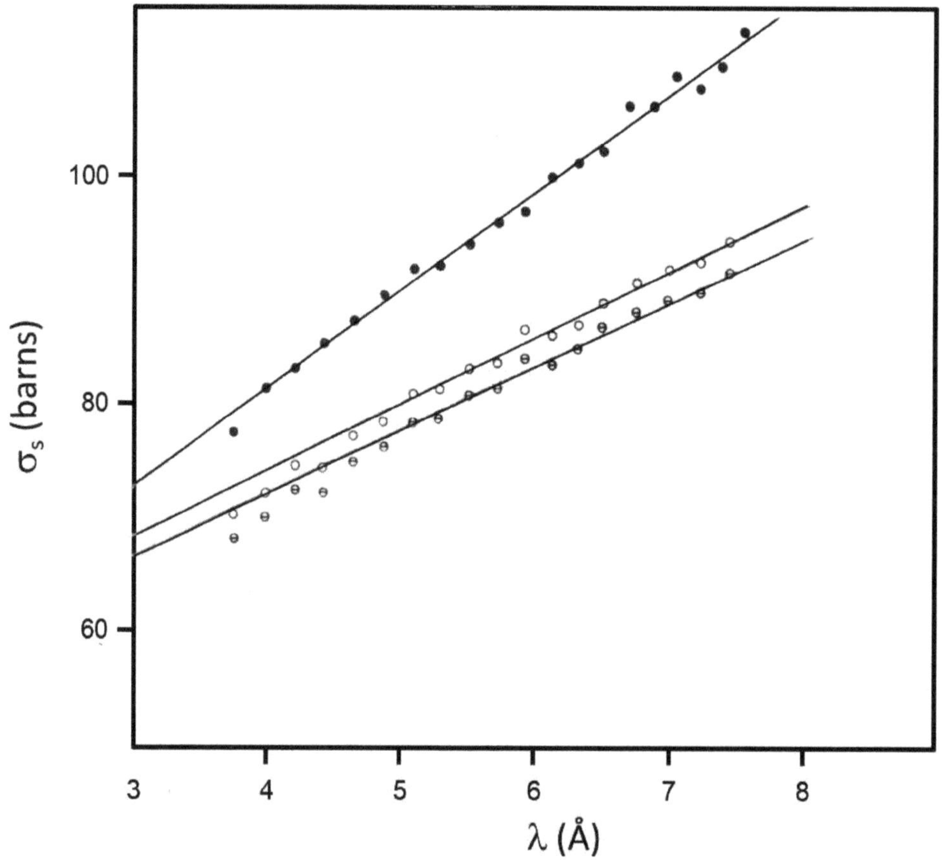

Figure 2. Total scattering cross-section per H atom, σ_S/H, for tert-butanol, as a function of the neutron wavelength λ. Results for the liquid (upper) and for the two limiting values of the solid state. Straight line adjustment is well defined over the whole λ interval in the liquid state, and for λ > 5 Å in the solid state.

The more detailed study of the inelastic cold neutron scattering (INS) technique is useful, especially as a complementary technique to infrared or Raman spectroscopy. Oscillations due to internal rotations are, in general, not observed in spectroscopy, since they induce small variations in dipole moment, besides having low frequencies, usually in the microwave region, and being accessible in a gaseous state. By INS, it is possible to study both the energy levels inside the potential well and the relaxation processes from jumps and molecular reorientations. In the results obtained with INS in the study of tert-butanol [31], a frequency spectrum as a sum of seven Gaussian functions fitted the measured time-of-flight distribution and allowed assignment of peak positions. A barrier V = 4.0 ± 0.2 kcal/mol for CH_3 internal rotation was obtained, in agreement with the result from transmission, within the range of statistical errors. Quasi-elastic line broadening and Debye–Waller factors were analyzed in terms of models for molecular diffusion, and the results were compared with NMR data. It was concluded that cooperative rotational diffusion occurs in both solid and liquid states.

4. Self-Assembly and Lyotropic Liquid Crystals

Research on lyotropic complex systems (water/amphiphile/additives) requires a very large amount of previous knowledge of physics and chemistry. Focus on experimental concepts is necessary, besides some knowledge on theories. A purely theoretical approach is still unable to deal with the very complex multicomponent aqueous solutions to be discussed now.

Starting from physics, a good background in the states of matter and their phase transitions is required—of concepts of order and disorder in structures of crystals and polycrystalline, and of phase diagrams as a function not only of pressure and temperature (as for a single component) but also of relative concentration in multi-component systems, as in metallurgy.

Starting from chemistry, the focus must be on water and its anomalies, the structure of the water molecule and the effects of the H bond, ionization of water and pH, hydrophobic and hydrophilic effects, formation of micelle aggregates by self-assembly, critical micelle concentration, behavior of soaps and detergents in water solution, and phase diagrams as a function of relative concentration and temperature.

Figure 3 provides some idea about micelles, the transient aggregates formed in equilibrium with monomers in solution, changing forms with concentration and temperature. At the CMC, single molecules form spherical micelles in the isotropic I phase; with increasing concentration, hexagonal H_α and lamellar L_α phases may form at temperatures above the Krafft line that separates these structures from the gel and coagel crystalline phases. Complex cubic structures may form between the H_α and L_α phases.

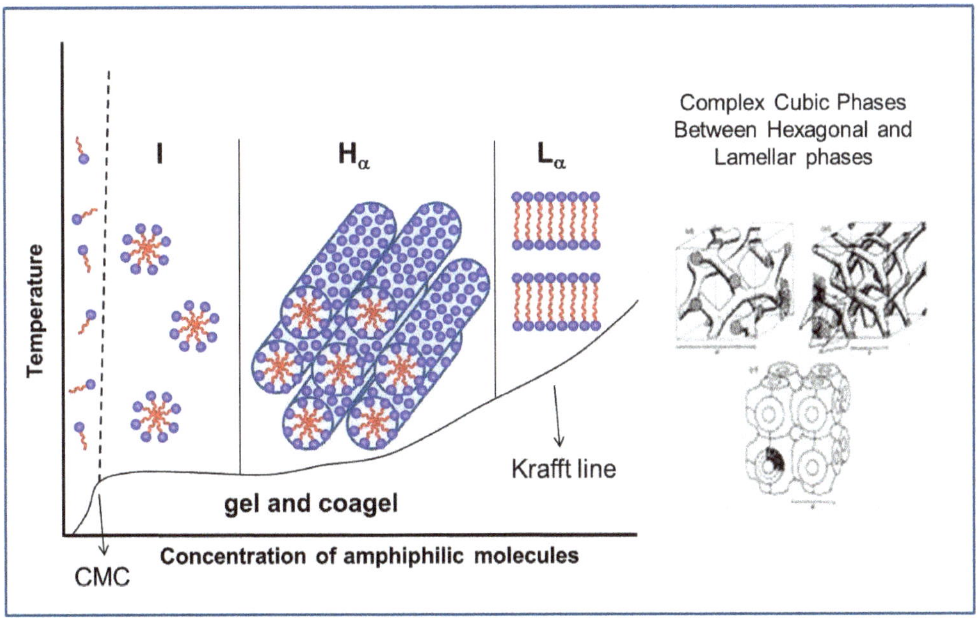

Figure 3. Scheme of phase diagram as a function of concentration and temperature for a lyotropic aqueous system. At the CMC, single molecules form spherical micelles in the isotropic I phase; with increasing concentration, hexagonal Hα and lamellar Lα phases may form at temperatures above the Krafft line that separates these structures from the gel and coagel crystalline phases. Complex cubic structures may form between the H_α and L_α phases.

An important point to emphasize is that the hydrocarbon chains (HC) inside the aggregates in lyotropic liquid crystals (LLC) are in a disordered α state above the Krafft temperature, while they are in extended β conformation below that temperature. The state of the HC chains can be determined from X-ray diffraction, from the peak corresponding to the average distance between the HC chains: diffuse at 4.5 Å in α state and sharp at 4.1 Å in β state.

4.1. The Beginning of the CrysLab IFUSP

Due to the lack of technicians in the group, it was necessary to "became an experimentalist" in order to assemble the electro-hydraulic infrastructure of the lab with diffractometers. It was also necessary to understand the basic physics of the interaction of X-rays with matter.

The technique of X-rays used in the CrysLab corresponded to the usual elastic coherent scattering, where the incident low-energy X-ray photon with wavelength λ induces electron vibration in the same frequency, and the vibrating charged electron emits a photon, in a scattering angle 2θ in relation to the incident photon. The momentum transfer is measured in terms of the scattering vector modulus $q = 4\pi \sin(\theta)/\lambda$, and the intensity is recorded by photographic technique or by electronic detector. The scattering vector has a defined direction in space. The interaction occurs with the region of the sample illuminated by the X-ray beam, and during the whole exposition time.

The measured intensity can be written as $S(q) = P(q) \times S(q)$, where $P(q)$ is the form factor of the individual scatter, and $S(q)$ is the structure factor of the scatter ensemble. These functions may be continuous or present peaks, corresponding to distances $d = 2\pi/q$. The term "X-ray diffraction" is used when the sample is a crystal with a well-defined structure and oriented Brag peaks related to the translational symmetry. When the individual scatter is a "particle", $S(q)$ refers to the ensemble of particles. The term "small angle X-ray scattering" (SAXS) is used when the units are large enough to require small 2θ values. In view of the size of the lyotropic structures, the CrysLab project required both SAXS and diffraction.

For liquid crystals, the situation requires careful analysis of X-ray results, since the scattering shows, in general, few peaks that may depend on the sample orientation, observed by polarized optical microscopy (POM).

The first paper of the X-ray CrysLab of IFUSP was in the field of LLC, and was directly connected to the seminal work by the Canadian chemist who found the two uniaxial nematic lyotropic phases [32]. It was with a phase Nd of sodium decyl sulfate, and it was also the first international article on the structure of a nematic LLC [33]. The initial samples (and publications) were made using the recipe of the chemists, adding decanol and/or salt to only two detergents, which were difficult to synthesize or buy (sodium decyl sulfate and potassium laurate), and diffraction was detected by bidimensional film images.

4.2. Introduction to Physical Chemistry

Working full time at IFUSP after tenure was hard. It was difficult to reconcile my personal life with teaching and the academic bureaucracy, in addition to research in the CrysLab. My introduction to nematic LLC was due to a "chance event". I was not, in fact, fully prepared for the difficulties of the field. However, I did my best to fulfill the tasks. Articles were published, one master's degree was finished, one Ph.D. in LLC was completed, and I started to work hard to get the Dozent frei (DF) academic degree (as exists in Germany), mandatory for a scientific career at IFUSP.

When a biaxial nematic lyotropic phase was discovered experimentally in the USA [34], several groups of physicists entered the field. However, the ternary system is too complex for physicists, and it became clear that the available theories were unable to explain the biaxial phase that is intermediate between the Nc uniaxial phases (cylindrical micelles) and the Nd uniaxial phases (discotic micelles).

I then realized that I really needed to engage with physical chemistry by myself in order to proceed in LLC. I then studied the DLVO theory on forces between charged surfaces interacting through a liquid medium. I focused on the planar geometry with results in the Nd phase—passing the Nc phase (with potassium laurate) to the Ph.D. student—and the work on interactions between micelles, considering Van der Waals attraction, Coulomb repulsion, and excluded volume gave interesting results [35]. A following master's degree studied the interaction between surface and magnetic orientation [36].

For my DF thesis, I worked alone to understand the process of magnetic orientation of the micelles due to the HC chains [37], the structure of the gel phase with Bragg peaks at lower temperature [38], and how to use other techniques (thermal analysis and electron microscopy) in LLC [39].

Over two decades (graduation in 1962, Ph.D in 1972, DF in 1982), the student with mathematical rigor turned into a "multitasks" scientist at the interface of physics/chemistry.

This first decade of the CrysLab ended in 1985, when I became Associate Professor at IFUSP, after participating in several international congresses. At that time, the CrysLab had already several published papers and formed two master's and two Ph.D. students in LLC. The last student focused on elastic constants, obtained from measurements of the critical magnetic field for the Freederickzs transition. It became clear, however, that the lab needed to be renewed, the old hand-control required automation, and a broader direction of research needed to be defined.

4.3. Micelles and Interface Curvature

A parallel didactic explanation is now necessary. "Amphiphilic" refers to a molecule made of two distinct covalently bonded components with different affinities for the solvent. One part possesses a high affinity for polar solvents (such as water), and another part has a strong affinity for nonpolar solvents (such as hydrocarbons). Examples of amphiphiles are molecules with hydrophilic and hydrophobic parts, with a paraffin tail and a polar head, such as detergents and some lipids. Grease or oil are not dissolved by water due to their opposite polarity, but are dissolved by aqueous solutions of amphiphiles. A micelle is an aggregate of amphiphile molecules formed spontaneously at a critical micellar concentration (CMC), observed experimentally by a drop in the superficial tension at the interface between the aqueous solution of amphiphiles and air. Micelles are formed inside the solution when monomers fill the interface with air, and are in chemical equilibrium with monomers inside the solution. Surfactants (surface active agents), known as soaps, are amphiphiles that have been used for cleaning purposes since antiquity.

It is possible now to enter in some aspects of interface curvature. The changes in the form of micellar structures require new concepts, and these came in 1976 through the concept of the molecular packing parameter, associating the molecular geometry with the polar–apolar interface curvature of the micellar aggregates. Defining **V** and $\boldsymbol{\ell}$ as the volume and the length of the paraffin tail and **a** as the surface area per molecule, at the interface, the "natural" parameter is defined by $\mathbf{po = V/a\ \ell}$.

The geometric definition gives a rough idea of the expected curvature of the surfactant aggregate (<1/3 for sphere, 1 for planar bilayer, >1 for inverted micelles). It must, however, be stressed that this parameter by itself does not solve the problem of the form of micellar aggregates, since other factors, such as amount of water bound to the polar head, may drastically change its actual value in a given aggregate. It is out of the scope of this review to enter into the problem of curvature free energy, a problem solved for membranes in vesicle geometry, considering bending rigidity and Gaussian curvature.

However, something about this issue is mentioned in Section 4.5.

4.4. Broader Research Directions of the CrysLab IFUSP

Detailed characterization of nematic phases with both optical and X-ray techniques were developed, and automation of the electronics of all the equipment, required much focus on the experimental aspects of the research. At the same time, the research lines opened in the 1990s in the more general direction of "non-crystalline materials", and new techniques were used, together with new collaborations, also at the international level.

Regarding LLC, it became clear that it was necessary to discover the lyotropic nematic domain in another system of broader interest, and the choice was to focus on sodium dodecyl sulfate (SDS), also named sodium lauryl sulfate (SLS), chemical formula $NaC_{12}H_{25}SO_4$, the standard amphiphile used for micellar systems in physical chemistry, which is easy to buy. It is a simple hydrocarbon chain (n = 12), with a terminal CH_3 and a polar head, as displayed in Figure 4.

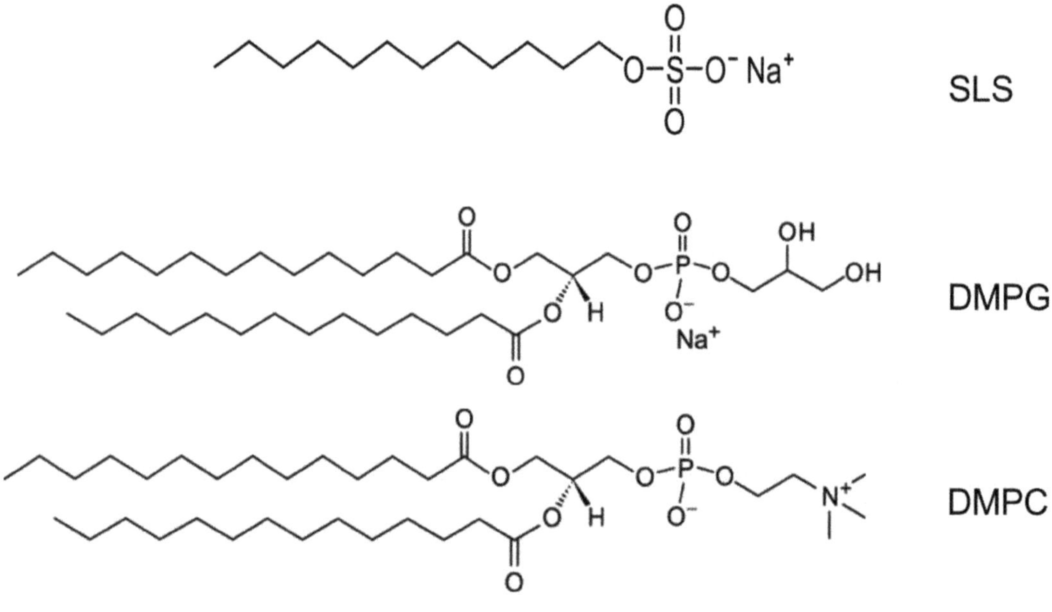

Figure 4. Molecules studied: detergent SLS (sodium dodecyl sulphate) with one HC chain, and phospholipids with two HC chains, charged DMPG and neutral DMPC (item 5.1).

With some effort, using POM and surface effects in capillaries, it was possible to discover (by trial and error) an N_C phase at 23 °C with weight composition 25.00% SLS/70.53% H_2O/4.47% decanol [40]. A systematic search of the nematic domain with SLS was then made as a function of two concentration variables, the water/amphiphile (Mw) and the decanol/amphiphile (Md) relative molar ratios, with good results [41]. The ternary phase diagram at room temperature showed the nematic domain between the isotropic I, hexagonal H_α, and lamellar L_α phases already studied by chemists. The phase diagram as a function of temperature and Md indicated a change in micellar form at the transition Nd–Nc, for Md = 0.38, with Mw varying in the interval 40–45. Furthermore, a comparison with sodium decyl sulfate and K laurate in terms of the new variables Mw and Md (Table II of [41]) revealed that the transition between the two uniaxial forms occurred for very near Md values, while Mw had a large variation (from 21 to 43), depending on the size of the micellar aggregate. These original results defined new directions for the research.

The diversification of directions of research included a master's degree on the isotropic phase of the binary system SLS/water, by SAXS with isotropic one-dimensional analysis by electronic detector. For concentrations up to 10 wt%, a constant peak was observed, corresponding to 36 Å, due to the inner structure of the micelle, compatible with light scattering results, giving a hydrophobic radius of 16.7 Å (extended chain).

The diversification included also a doctorate in binary amorphous alloys and more doctorates using several techniques. More students have been formed, and several papers have been published, but the focus here turns now to some specific problems that drew my attention over a long period, trying to join basic physical concepts with the behavior of these very complex systems.

From the discovery of the nematic domain with SLS, studies went in the direction of a detailed exploration on the changes in the micelles at the phase sequence isotropic (I)–Hexagonal (H_α) in the binary system SLS/water. The basic discussion was on the micellar radius, known to be that of the extended chain in spherical micelles in I phase. In the H_α phase, the distance between cylinders is given by X-ray diffraction, but the cylinder length is unknown, and it was usual to deduce the cylinder radius from the hypothesis of "infinite cylinders", imposing homogeneous distribution of amphiphile and water in the plane perpendicular to the cylinders. This usually resulted in a cylinder paraffinic radius smaller than the extended length of the chain. The alternative was to admit that the cylinders are finite, with polydispersity in length, but keeping the same radius of the spherical micelle.

This possibility was checked analyzing the variation of the hexagonal peaks with SLS concentration, compared with the cylindrical form factor calculated assuming a two-step radial distribution function (inner paraffinic region and an outer shell containing polar heads and hydration water) [42]. This means that the basic principle that matter density is the same at the atomic level and at the macroscopic level was extended to the transition I–H_α. The analysis considered the relations between volumes occupied by micelles and water and some possible forms for the micelles, together with possibilities of both two-dimensional and three-dimensional short-range ordering. The results indicated [42] that in the concentrated I phase, micelles have stable dimensions and are slightly anisometric, with micellar growth at the I–H_α transition, up to 230 Å in the H_α phase.

From the new results with SLS, another basic question started to deserve further investigation, namely the direct Nc–Nd transition in the SLS system, without an intermediate biaxial phase [43].

The growth of micellar cylinders in SLS/water was further studied by analyzing the behavior of the hexagonal cell parameter $\boldsymbol{a}$ in the function of volume concentration c_v along the phase sequence transitions I–H_α–M_α (monoclinic distorted hexagonal). A functional behavior $\boldsymbol{a} \propto c_v{}^{-1/3}$ was obtained [44], indicating tridimensional expansion, that is, finite length, instead of the coefficient $-1/2$ expected for only in-plane separation.

At the same time, a new approach relating the micellar bending energy to the surfactant packing parameter for SLS indicated transformation of prolate ellipsoid into a spherocylinder for anisometry ν > 1.8 [45], a possible explanation for the transition I–H_α. The subsequent question, related to the role of decanol in these lyotropic aggregates, is being investigated still, without a complete answer. This direction started with analysis of the sequence I–Nc–H_α in the ternary system SLS/water/decanol, after the determination of the nematic domain [41].

Instead of describing all the steps of the research done throughout more than three decades on this problem (including 3 Ph.D. these and many papers), it is more efficient to make reference to an invited paper published in the volume of Liquid Crystals in honor of Saupe, discussing the up-to-now unsolved problem of the micelles in the biaxial nematic phase [46], with 91 references.

However, the present state of the problem focusing on theoretical models is discussed in the next section.

In the context of this section, it is still worthwhile to mention that a detailed structural study of the H_α phase in SLS/water was performed, and the electron density maps obtained from the X-ray diffracted intensities allowed determination of the micellar parameters [47].

Regarding expertise in X-ray crystallography, the collaboration with the Italian group of Ancona began in the second decade with a further advance regarding the analysis of complex systems [48].

From the point of view of applications, lyotropic liquid crystals are now used as templates for a variety of nanostructures; in particular, the nematic Nc phase of SLS is used as templates for carbon nanotubes [49,50].

It is worthwhile also to mention that the nematic lyotropic system sodium dodecyl sulfate, decanol, and water was studied by rheology and deuteron nuclear magnetic resonance NMR measurements [51], showing strong evidence that the N_C and Nd nematics are textured nematics of the flow-aligning type, and not of the well-established tumbling-type surfactant nematics.

4.5. The Nc-(Biaxial)-Nd Transition

When the existence of the two uniaxial nematic phases was discovered by chemists [32], they were classified as Type I and Type II, depending on their orientation on the magnetic field by NMR measurements on samples of detergent/deuterium/additives (decanol and /or salt), with positive and negative bulk diamagnetic anisotropy. A next paper was published [52] with a table summarizing all the phases known at that time, which already defined Type I as cylindrical (CM) and Type II as discotic (DM) micelles, and their relationship to the parent "hexagonal" and "lamellar" lyotropic liquid crystals. They also had different surface orientations, so that it was very clear that micelles changed symmetry with changes in composition. This was five months before the paper on the discovery of the biaxial phase [34], intermediate between the two uniaxial phases in a specific ternary system (with potassium laurate), investigated by temperature variation in a specific sample composition, related to the entrance of physicists into this field.

Since the beginning, it was clear that there was a change in the symmetry of the micellar object, and that this could happen easily only in a lyotropic system, since thermotropic LC did not present a change in molecular symmetry. As already discussed in Section 4.4, and in the published review [46], the problem had no trivial solution. However, it is worthwhile to discuss in some detail the theoretical models that effectively approached experimental results.

A first attempt to model the symmetry transition, in collaboration with a theoretical Italian chemist continuing a work on the I–H_α transition [45], was an adaptation of the bending elastic theory of bilayers in vesicles to the micellar case, in terms of the surfactant parameter (already mentioned in item 4.3). The nematic cylindrical (Nc)-nematic discotic (Nd) phase transitions are correlated with a change of micellar form from spherocylinder (SC) to square tablet (ST) [53], which occurs geometrically in a continuous way, with an intermediate biaxial object. Good agreement was obtained for three amphiphile/decanol/water systems, where the transition occurs as a function of the decanol/amphiphile molecular ratio.

This result inspired a Maier–Saupe model for a polydisperse solution of micelles of axial symmetry [54], giving a phase diagram with a biaxial phase at small average anisometries and finite dispersion in size. A further model [55] considered, with a statistical microscopic approach, a mixture of cylinder and disk micelles, with changes of uniaxial micellar form occurring either smoothly or abruptly, to mimic the possibilities of both the biaxial phase and coexistence.

In another direction, Mukherjee in India [56] described new topologies in the phase diagrams involving biaxial nematic liquid crystals, finding a direct isotropic–biaxial nematic phase transition and three different biaxial nematic phases, outlining how the novel phase diagrams could be detected experimentally.

Subsequent models with exact statistical mechanics calculations for a Maier–Saupe lattice model considered the inclusion of extra degrees of freedom to mimic a mixture of discs and cylinders. A quenched distribution of shapes leads to a phase diagram with two uniaxial and a biaxial nematic structure. A thermalized distribution, however, precludes the stability of this biaxial phase. The introduction of a two-temperature formalism (to mimic a separation of relaxation times) shows that a partial degree of annealing is already sufficient to stabilize a biaxial nematic structure [57,58].

Biaxiality could be the result of perpendicular alignment of uniaxial particles of cylinder-like and disc-like geometry in a mixture. A model for a general distribution of micellar anisometries could fit the original experimental data of Yu and Saupe's well-known 1980 paper [34], yielding a bimodal distribution, with the presence of two quadrupoles referred to as objects of opposite symmetry [59], giving support to the rationalization of the biaxial phase for lyotropic systems in terms of a polydisperse mixture of rod-like and disc-like micelles.

Recent results by Mukherjee investigate the influence of an external magnetic field [60]. The possibility of the various phases, including the coexistence phases, is explored by means of variation of the concentration [61]. The model describes the first theoretical observation of the phase transition between two biaxial nematic phases. The theoretical predictions are found to be in good qualitative agreement with available experimental results.

Regarding experimental results, it is worth mentioning a very recent paper [62] with evidence of positive and negative biaxial phases, with a transition point determined by means of optical image processing, and a careful discussion of the available literature.

After the above discussion on lyotropic systems with symmetry transitions, it is also worthwhile to mention the possible thermotropic biaxial phase, which has been sought for already more than 40 years. Even before this, a biaxial phase was predicted theoretically, without mention of lyotropics.

Here, it is worthwhile to focus on some articles from this period.

First of all, a paper by Chadrasekhar [63] on optical studies carried out on a nematogenic copper complex, which incorporates the features of both rod-like and disk-like molecules, showing both a biaxial phase in the pure complex and a uniaxial–biaxial transition in binary mixtures with temperature variation.

In another interesting paper, measurements were carried out on two liquid crystalline organosiloxane tetrapods [64], and results unambiguously showed the existence of a biaxial nematic phase below a uniaxial nematic phase.

A very recent paper considered colloidal particles suspended in liquid crystals [65], which exhibit various effective anisotropic interactions that can be tuned and utilized in self-assembly processes.

So many decades of investment clearly show that the solution is not a strict focus on "mathematical exactness" with no connection to experimental reality.

5. Biological and Bio-Inspired Liquid Crystals

Biology is not an exact science. Studies on life have existed since antiquity in the ancestral practices of medicine, and Aristotle started systematization by comparison, classifying plants and animals, and separating animals those with blood and without blood. Only in the 17th century, based on careful observation of gardens since infancy, the Swedish scientist Carl Linnaeus proposed a more exact systematization in his work *Systema Naturae*, published in Latin in 1735, which is origin of the modern taxonomy. He used the word "biologie" in Latin, and the German translation of his book in 1771 also used "biologie".

The invention of the microscope in Holland in the 17th century made possible the observation of bacteria and sperm, and with the advance of better microscopes, cells could be identified. Biology became a science at the turn of the 20th century, recognizing the cell as the basic unity of life, genes as the basic units of heredity, and evolution as the principle for changes in life.

Molecular biology starts with the determination of the helical structure of DNA, by X-ray diffraction, and attribution of nucleotides as its basic molecular structure. The history of the union of molecular biology with genetics and biochemistry can be followed in [66].

Molecular biology approaches exact science only when samples have an exact chemical composition, since only then can the analysis follow exact principles. Biological material does not in general satisfy such criteria. Academic work in biochemistry and biophysics is usually made on biomimetic systems, such that the effect of the components can be rationally studied in analytic form.

5.1. DNA and Lipids

Biology has always interested me in life. Its lack of exactness pushed me away during graduation, but attracted me afterwards. During the period of the reactor, I participated in studies of DNA samples by slow neutron transmission, at room temperature, comparing a dry sample and a wet sample with 7.8% moisture. The total neutron cross-section of the water was obtained, and the only conclusion was that it behaved as free water. After becoming the first woman Full Professor in IFUSP (1991), I was engaged in the creation of the Department of Applied Physics in IFUSP.

In the period of broader research directions (Section 4.2), a collaborative project with the laboratory at Ancona, Italy, started with LLC and continued in these systems of biological interest. A published article was written with the participation of a Brazilian student in Ancona on effect of a drug in a biomimetic membrane [67]. This was followed by a work with an Italian student in São Paulo [68], related to the same problem of the structure of the H_α phase, already discussed for SLS. This exchange of subjects of research lasted intermittently for almost 30 years, depending on mutual visits and interests.

A deoxyguanosine–water system, presenting an intermediate cholesteric Ch phase between isotropic I and hexagonal H_α phases, showed a behavior typical of infinite or long flexible cylinders in H_α phase, correlating the hexagonal parameter $\boldsymbol{a}$ with the lipid volume concentration c_v, according to $\boldsymbol{a} \propto {c_v}^{-1/2}$ [68].

After this result, studies of micellar growth were extended from SLS to lipids, focusing on transitions to cubic lyotropic phases (Q), with measurements made in Europe [69,70]. There is much interest in the bicontinuous cubic phases with intricate structures, as sketched in Figure 3, both from the academic point of view and based on their possible existence in biological systems. Cubic phases with spherical micelles also exist between the I and H_α phases, which is also an intricate problem that drew my attention.

The lipid OLPC (oleoyl-lyso-phosphatidyl choline), presenting a cubic bicontinuous phase (Q^{230}), has the phase sequence I–H_α–Q–L_α. The behavior measured in the H_α phase gave $\boldsymbol{a} \propto {c_v}^{-x}$, with $x <$

1/3 (in fact, x = 0.226), indicating that micelles grow from spherical in the I $-H_\alpha$ phase transition to infinite in the transition H_α–Q^{230} [69]. It was possible to define a function representing the cylinder growth in the function of c_v. Results with SLS/water and SLS/water/decanol could also be compared with statistical mechanical calculations for systems with self-association.

After that, three lipid–water systems (PLPC, OLPC, and DTAC) with transitions to cubic phases following the sequence I–H_α–Q, either bicontinuous (Q^{230}) or cubic micellar (Q^{223}), received a more complete X-ray study [70]. PLPC (palmitoyl-lyso-phosphatidyl choline) has the phase sequence I–Q^{223}–H_α–L_α, while DTAC (dodecyl–trimethyl-ammonium chloride) has the phase sequence I–Q^{223}–H_α–Q^{230}–L_α. The behavior of the hexagonal parameter and the electron density map were obtained, showing epitaxial relationships occurring at the H_α–Q transitions, in the direction of the cylinder growth. It was possible to propose a mechanism for the transformation of the structures [70].

This type of study continued with another Ph.D. in our lab, focusing on the transition I–Q^{223} in PLPC from the isotropic side [71]. The more diluted region gives the particle form factor P(q), and in the more concentrated region, the SAXS curve is fitted by the product of P(q) S(q), with the structure factor calculated using a hard sphere interaction potential. It is found that PLPC micelles remain with a prolate ellipsoidal shape of constant anisometry $\nu = 1.80 \pm 0.05$ in the whole concentration range in the I phase. It is concluded that the formation of the Q^{223} phase requires that micelles remain spheroidal, in contrast with SLS, where micelles grow to slightly higher anisometry ($\nu \sim 2.4$) before the I-H_α transition, in agreement with the elastic bending theory in terms of the surfactant parameter [45].

The collaboration with Italy extended to organic chemistry in Bologna, in a study with DNA carried out in our X-ray lab with samples from Italy [72]. Aqueous solutions of fragmented DNA were studied by SAXS, varying the concentration. The less concentrated solution (I phase) gives the DNA form factor, in good agreement with the B form of DNA. The semidilute regime until the I-cholesteric phase transition has an interference peak position that behaves with concentration with the exponent ½ for an effective rod length of L = 340 Å. The structure factor is obtained by dividing the SAXS curves by the form factor, and is modeled with Gaussian functions, with peak broadening varying also with the exponent ½, indicating a short-range order slightly above first neighbors.

The analysis of the short-range order of rodlike polyelectrolytes in terms of changes from exponent 1/3 to 1/2 when going from a dilute to semidilute regime (after a critical concentration) was extended in the following paper [73] to virus solutions taken from the literature. The number of ordered next neighbors arranged locally around a central particle indicates also only first neighbors, in good agreement with Monte Carlo results, when the ionic strength of the solution is low.

The transition I–Q^{223} in PLPC was further investigated in our laboratory, now from the side of the cubic micellar phase [74]. It was found that a local cubic order already exists in the concentrated I phase, but the positional correlation is between clusters with 5 micelles. In the lower concentration limit of the Q^{223} phase, the long-range order is established, but micelles have the same paraffinic volume as in the I phase.

5.2. A Special Biomembrane

The next and final subject that drew my attention, with long-lasting research, was a special negatively charged phospholipid, with unusual phase transitions from the gel phase to the liquid crystalline phase. Figure 5 shows a scheme of the order–disorder phase transition of phospholipids in membranes, showing the separation between the high-concentration region, where LLCs are studied, and the low-concentration region, where vesicles are studied.

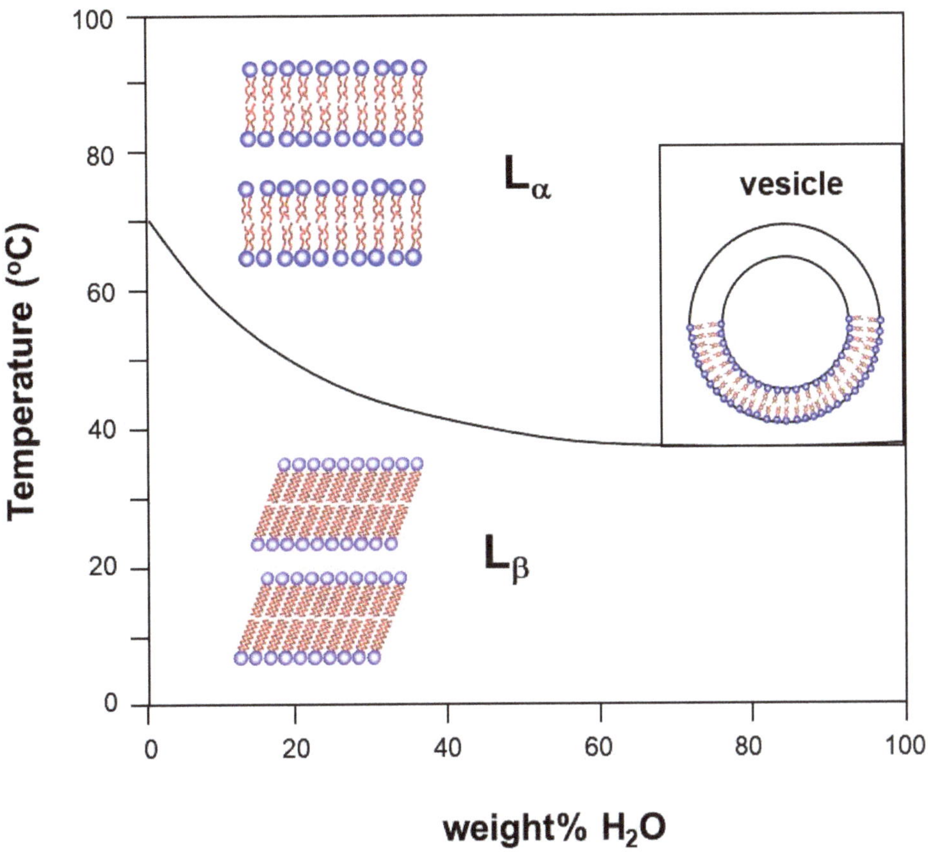

Figure 5. Phase diagram of usual phospholipid membranes, with order (Lamellar L_β)–disorder (Lamellar L_α) transition at the Krafft line, in the low-water-content region. Vesicles (unilamellar or multilamellar) are formed in the high-water-content region.

Phosphatidylglycerol (DMPG, 1,2-dimyristoyl-sn-glycero-3-[phospho-rac-glycerol]) is the most abundant anionic phospholipid present in prokaryotic cell membranes and has been extensively studied as a model for negative membranes, in contrast to the neutral phosphatidylcholine (DMPC, 1,2-dimyristoyl-sn-glycero-3-[phosphocholine]). Figure 4 shows these phospholipid molecules, with two HC chains, compared with SLS, with one HC chain.

DMPG was already being studied by the biophysics group at IFUSP at low concentrations, as usual (1 to 10 mM in HEPES buffer), by several techniques. The main transition showed, by light scattering, three different regions. At low (20 °C) and high temperatures (30 °C), the sample is turbid, whereas between the transitions, a transparent and viscous phase is observed. DMPG does not exist in eukaryotic cells, and my interest was in trying to correlate this absence with its unusual thermal behavior.

I joined the team in structural studies using X-rays, which required higher concentrations (50 mM) for reasonable intensities. The initial focus on this subject resulted in a series of four articles in collaboration with the group at IFUSP, using the Brazilian synchrotron, with SAXS curves obtained with a linear position-sensitive detector [75–78].

The initial SAXS results [75] showed for DMPC the Bragg peak typical of multilamellar vesicles, with a repeat distance of 66 Å, while for DMPG, it was only a broad peak, indicating single bilayers, for the range of temperatures studied (10–45 °C). The peak position changed continuously in the intermediate region, while the intensity showed an unusual effect, with a decrease along the intermediate region and a minimum at its end, and bilayer electron density profiles fitted to the SAXS curves.

The light scattering thermal profile was similar to that yielded by the 10 mM DMPG dispersion, with only slight change in the temperature interval. In differential scanning calorimetry (DSC), DMPC presents a pretransition and a very sharp and strong melting peak at Tm = 23.5 °C. In contrast, a complex calorimetric profile is observed for 50 mM DMPG at low ionic strength, between 10 °C and 30 °C, besides the pretransition. At high ionic strength, this unusual thermal behavior disappears. The conclusion [75] was for a DMPG complex melting regime at low ionic strengths, between re-named temperatures T_m^{on} and T_m^{off}, the onset and outset of the melting regime, respectively.

The next step was to decrease the minimum scattering angle measured in SAXS, in order to reach larger correlation distances. Besides the bilayer peak present in all phases, a peak corresponding to a mesoscopic structure at 400 Å was detected for DMPG only in the intermediate region [76]. The behavior of this repeat distance d was analyzed as a function of the lipid concentration c_v. For a bilayer, the symmetry decouples the three-dimensional volume partition of the lipid in a one-dimensional factor (perpendicular to the bilayer) and a two-dimensional factor (in the bilayer plane). For infinite planar bilayers with thickness t, c_v coincides with the linear fraction of lipid occupancy (t/d). When the lamellae have fluctuations or are finite, it is possible to define a lipid surface fraction (LSF) such that $c_v \propto \mathrm{LSF}(t/d)$. The value LSF can be obtained from knowledge of c_v, t, and d, and it is <1, as shown in Figure 5 from [76], which is compatible with in-plane and not lamellar correlation.

In the same article, a new technique was used in the post doctorate of a Brazilian student in Germany, with observation of giant DMPG vesicles with phase contrast light microscopy, showing that vesicles "disappear" upon cooling below T_m^{off} and "reappear" after reheating above T_m^{on} [76]. It was proposed in this paper that the melting regime corresponds to unilamellar vesicles with perforations, with possible biological relevance.

The third paper in this series dealt with the effect of salt addition, with increase in ionic strength investigated by SAXS and optical microscopy (OM), indicating the changing of vesicles [77]. By SAXS, the broad bilayer peak arising from the electron density contrasts within the bilayer was observed at all temperatures up to 100 mM NaCl addition. At higher ionic strength (250–500 mM NaCl), an incipient lamellar repeat distance around d = 90–100 Å is detected, superimposed to the bilayer form factor, indicating a loose multi-lamellar order of only four bilayers, accounted for by DLVO theory [77].

The final paper in this series used DSC, turbidity, and OM of giant vesicles, combining phase and fluorescence microscopy to study the narrowing of the transition region with the increase of ionic strength [78]. It was possible to correlate the complex DSC profile with the vanishing of the bilayer optical contrast. The indication is that bilayers are perforated along the transition and the bilayer completely loses the optical contrast, directly related to turbidity.

The study of DMPG continued within the collaboration with the Ancona group, and my interest was in filling the gap in Figure 5 at intermediate concentrations. Another long-lasting series of three articles used new data at higher DMPG concentrations, obtained in the Brazilian synchrotron with SAXS and WAXS (wide angle for detection of melting of HC chains) and also POM [79–81]. The sodium salt of DMPG, and bidistilled water, were used to prepare samples with concentrations in the interval of 70–300 mM. The same buffer (10 mM HEPES pH 7.4 with 2 mM NaCl) was used throughout, without any further adjustment of pH of the lipid dispersion, to ensure the same additional low ionic strength as in all the previously published results on DMPG at smaller concentrations. However, the measured pH remained always above the apparent pK of DMPG, ensuring that DMPG can be assumed to be fully deprotonated and that the effects are due to changes in DMPG concentration. A thermal bath was used for temperature variation from 12 °C to 55 °C.

At these higher concentrations, the melting regime persists, but it is not transparent. Defined SAXS peaks appear, and a new lamellar phase Lp with pores is proposed to exist above 70 mM DMPG, starting at 23 °C (3 °C above T_m^{on}) and losing correlation after T_m^{off}. A preliminary qualitative analysis of such a large amount of data led us to focus on SAXS curves with 150 mM DMPG to analyze the very complex behavior of the mesoscopic correlation peak with temperature [79]. The lipid surface fraction given in Figure 3 of [79] shows a near match of the previous results up to 70 mM in [76], and the new results up to 300 mM, providing evidence that Lp is not a normal bulk lamellar phase. In the sample with 150 mM, the DSC showed a clear peak corresponding to the new Lp phase, WAXS showed the continuous melting of the HC chains, and the intensity of the broad bilayer peak had a complex behavior. However, a pore model was developed and gave a good fit to the 70 mM curve in the intermediate region [79].

These data were analyzed in another paper, presented in an SAXS conference [80], using a model of water-penetrated bilayers (instead of pores) but without good agreement with the details of the variation of the intensity of the broad bilayer band.

Only six years later was the final paper published, with a detailed pore model fit to the experimental results [81]. Large and small toroidal pores are necessary to explain the SAXS results. Pores have DMPG in the fluid conformation, whereas the flat region of the bilayer has DMPG molecules in fluid and in gel conformations. Electron densities consider all molecular and ionic species that characterize the system and the temperature dependency of their volumes. The gel phase transforms initially, at 19.4 °C, in uncoupled bilayers with large pores, which transform into small pores along the lamellar phase. The minimum intensity of the SAXS bilayer peak at 30 °C corresponds to a maximum number of small pores, and above 35 °C, the system enters into the normal lamellar fluid phase, without pores. The charge is estimated and shows that the regions with pores contain fewer Na^+ ions per polar head; hence, when they are forming, there is a release of Na^+ ions toward the bulk.

The opening of pores by temperature is certainly of biological importance, and may explain why DMPG cannot exist in eukaryotic cells. With this final article, I finished my involvement with molecular biophysics, but my interest in biology continues.

5.3. *Applications of Scientific Knowledge*

This review has discussed, after an introduction, the relative importance of theory and experiment—a question which has no trivial answer, since knowledge is constructed from both. Now, the discussion is on the importance of applications of the acquired scientific knowledge, and some focus on the question of basic vs. applied science is in order. The search for knowledge in humans comes from innate curiosity, present from early childhood, and from the wish/necessity to understand the world around each one of us—not necessarily from material needs. Therefore, the drive for basic research is in general not the same as for applied research, and applications can develop only after basic knowledge has been acquired.

The next sections focused on my own trajectory, but inserted in a more general context of the study of condensed matter physics and soft matter, also approaching chemistry and biology. My aim is to link basic knowledge with innovative ideas.

I finish this section with some relevant interdisciplinary research taken from the literature, regarding interesting and novel applications in the direction of life. The comparison of basic properties of liposomes (spherical vesicle with lipid bilayers) and polymersomes (self-assembled block-copolymer vesicles) as cell-mimicking [82] allows a better a priori choice and design of vesicles. On the other hand, polymersomes have showed the ability to significantly enhance the efficacy of the antibiotics killing established intracellular pathogens and represent innovative applications for the eradication of intracellular bacteria [83].

A recent review on molecular bionics principles for supramolecular design [84] gives guidelines to design materials for biomedical applications, for tissue substitutes or in drug delivery vehicles, with applications in biomaterials engineering.

6. Conclusions

This review did not mention temporary collaborations—for instance, works on microemulsions or cationic micelles—nor all of my academic research work. The focus has been on some non-trivial subjects which required my special attention, leading to series of correlated papers, pursuing solutions, and searching for internal consistency. Experimental and theoretical techniques are tools for research, but the real problem is the correct understanding of basic principles, unifying different branches of knowledge.

My interest is in joining experimental results with simple theoretical approaches, which are able to illuminate and clarify unsolved and intriguing problems. A principle that emerged was the analysis of basic scaling laws' impact on distribution of matter, correlating structures not completely defined in the three dimensions of space with accessible and known macroscopic molecular densities.

I could also mention efforts in the direction of teaching and scientific divulgation, in the field of these complex systems [85], and also work throughout my whole academic career applying physical methods to the study of human physical evolution [86].

Funding: This research received no external funding.

Institutional Review Board Statement: Not applicable.

Informed Consent Statement: Not applicable.

Acknowledgments: Thanks are due to all the co-authors of my publications, in particular Karin Amaral Riske who, besides bringing to me the subject of the special phospholipid DMPG (Section 5.2), helped formatting the figures in this review. Thanks also for the financial support, both institutional and personal, from several agencies throughout my career, and in particular, FAPESP (of the state of São Paulo) and CNPq (of the Brazilian federal government).

Conflicts of Interest: The authors declare no conflict of interest.

References

1. Fultz, B. *Phase Transitions in Materials*, 2nd ed.; Cambridge University Press: Cambridge, UK, 2020. [CrossRef]
2. Tilley, R. *Crystals and Crystal Structures*; John Wiley & Sons: Hoboken, NJ, USA, 2020.
3. Menger, F.M. Supramolecular chemistry and self-assembly. *Proc. Natl. Acad. Sci. USA* **2002**, *99*, 4818–4822. [CrossRef]
4. Kralchevsky, P.; Miller, R.; Ravera, F. *Colloid and Interface Chemistry for Nanotechnology*, 1st ed.; Kralchevsky, P., Miller, R., Ravera, F., Eds.; CRC Press: Boca Raton, FL, USA, 2019; ISBN 9780367379759.
5. *Studies in Interface Science*; Elsevier: Amsterdam, The Netherlands, 1995.
6. Dunmur, D. 125 years of liquid crystals. *Liq. Cryst. Today* **2013**, *22*, 82–86. [CrossRef]
7. Gavroglu, K.; Simões, A. *Neither Physics nor Chemistry—A History of Quantum Chemistry*; MIT Press: Cambridge, MA, USA, 2012. [CrossRef]
8. Bell, E. The future of education in the molecular life sciences. *Nat. Rev. Mol. Cell Biol.* **2001**, *2*, 221–225. [CrossRef] [PubMed]
9. Jensen, W.B. The Origin of the Polymer Concept. *J. Chem. Educ.* **2008**, *88*, 624–625. [CrossRef]
10. Gaitzsch, J.; Delahaye, M.; Poma, A.; Prez, F.D.; Battaglia, G. Comparison of metal free polymer–dye conjugation strategies in protic solvents. *Polym. Chem.* **2016**, *7*, 3046–3055. [CrossRef]
11. Thorium in Brazil. Available online: http://www.aben.com.br/Arquivos/661/661.pdf (accessed on 11 July 2021).
12. Breit, G.; Wigner, E. Capture of Slow Neutrons. *Phys. Rev.* **1936**, *49*, 519. [CrossRef]
13. Fermi, E. Motion of neutrons in hydrogenous substances. *Ric. Sci.* **1936**, *7*, 13–52.
14. Zimmerman, R.L.; Amaral, L.Q.; Fulfaro, R.; Mattos, M.C.; Abreu, M.; Stasiulevicius, R. Neutron Cross Section of Pr, Yb, Lu, Er, Ho, and Tm. *Nucl. Phys. A* **1967**, *95*, 683–693. [CrossRef]
15. Van Hove, L. Correlations in Space and Time and Born Approximation Scattering in Systems of Interacting Particles. *Phys. Rev.* **1954**, *95*, 249–262. [CrossRef]
16. Egelstaff, P.A.; Pease, R.S. The design of cold neutron filters. *J. Sci. Instrum.* **1954**, *31*, 207. [CrossRef]
17. Larsson, K.E.; Bergstedt, L. Proton Motions in Complex Hydrogenous Liquids. I. A Cross Section for Quasi-Elastic Scattering of Slow Neutrons. *Phys. Rev.* **1966**, *151*, 117–125. [CrossRef]
18. Larsson, K.E.; Amaral, L.Q.D.; Ivanchev, N.; Ripeanu, S.; Bergstedt, L.; Dahlborg, U. Proton Motions in Complex Hydrogenous Liquids. II. Results Gained from Some Neutron-Scattering Experiments. *Phys. Rev.* **1966**, *151*, 126–132. [CrossRef]
19. Amaral, L.Q.; Vinhas, L.A.; Rodrigues, C.; Herdade, S.B. Certain Aspects of the Calibration and Resolution of Slow Neutron Spectrometers. *Nucl. Instrum. Methods* **1968**, *63*, 13–22. [CrossRef]
20. Allen, H.C., Jr.; Olson, W.B. Vibrational-rotational spectroscopy. *Annu. Rev. Phys. Chem.* **1962**, *13*, 221–240. [CrossRef]
21. Krieger, T.J.; Nelkin, M.S. Slow-Neutron Scattering by Molecules. *Phys. Rev.* **1957**, *106*, 290–295. [CrossRef]
22. Rush, J.J.; Taylor, T.I.; Havens, W.W., Jr. Proton Motions in Solids by Slow Neutron Scattering Cross Sections. *J. Chem. Phys.* **1961**, *35*, 2265–2266. [CrossRef]
23. Herdade, S.B. Slow neutron scattering and rotational freedom of methyl groups in several organic compounds. In Proceedings of the Symposium on Neutron Inelastic Scattering, Copenhagen, Denmark, 20–25 May 1968; IAEA: Vienna, Austria, 1968; Volume 2, pp. 197–204.

24. Rodrigues, C.; Vinhas, L.A.; Herdade, S.B.; Do Amaral, L.Q. Slow-neutron scattering cross-section for methanol, ethanol, propanol, iso-propanol, butanol, ethanediol and propanetriol. *J. Nucl. Energy* **1972**, *26*, 379–383. [CrossRef]
25. Rodrigues, C.; Amaral, L.Q.; Vinhas, L.A.; Herdade, S.B. Proton Motion in Methanol by Cold Neutron Scattering. *J. Chem. Phys.* **1972**, *56*, 3118–3121. [CrossRef]
26. Amaral, L.Q.; Vinhas, L.A.; Herdade, S.B. Methyl Rotation in Polydimethylsiloxane Studied by Neutron Transmission. *J. Polym. Sci. Polym. Phys. Ed.* **1976**, *14*, 1077–1085. [CrossRef]
27. Grapengeterl, H.-H.; Alefeld, B.; Kosfeld, R. An investigation of micro-brownian motions in polydimethylsiloxane by complementary incoherent-neutron-scattering and nuclear-magnetic-resonance experiments below room temperature. *Colloid Polym. Sci.* **1987**, *265*, 226–233. [CrossRef]
28. Wolfa, M.P.; Salieb-Beugelaara, G.B.; Hunzikera, P. PDMS with designer functionalities—Properties, modificationsstrategies, and applications. *Prog. Polym. Sci.* **2018**, *83*, 97–134. [CrossRef]
29. Amaral, L.Q. Study of the Atomic Movements of t-Butanol by Slow Neutron Scattering. Ph.D. Thesis, Institute of Physics, University of São Paulo, São Paulo, Brazil, 1972.
30. Amaral, L.Q.; Fulfaro, R.; Vinhas, L.A. Molecular Dynamics of Tert-Butanol Studied by Neutron Transmission. *J. Chem. Phys.* **1975**, *63*, 1312–1314. [CrossRef]
31. Amaral, L.Q.; Vinhas, L.A. Molecular Dynamics of Tert-Butanol Studied by Neutron Inelastic Scattering. *J. Chem. Phys.* **1978**, *68*, 5636–5642. [CrossRef]
32. Radley, K.; Reeves, L.W.; Tracey, A.S. Effect of Counterion Substitution on the Type and Nature of Nematic Lyotropic Phases from Nuclear Magnetic Resonance Studies. *J. Phys. Chem.* **1976**, *80*, 174–182. [CrossRef]
33. Amaral, L.Q.; Pimentel, C.A.; Tavares, M.R.; Vanin, J.A. Study of a Magnetically Oriented Lyotropic Mesophase. *J. Chem. Phys.* **1979**, *71*, 2940–2945. [CrossRef]
34. Yu, L.J.; Saupe, A. Observation of a Biaxial Nematic Phase in Potassium Laurate-1-Decanol-Water Mixtures. *Phys. Rev. Lett.* **1980**, *45*, 1000–1003. [CrossRef]
35. Amaral, L.Q.; Neto, A.M.F. Interactions between Micelles in Nematic Lyomesophases. *Mol. Cryst. Liq. Cryst.* **1983**, *98*, 285–297. [CrossRef]
36. Amaral, L.Q.; Rossi, W. Surface and Magnetic Orientation in a Type II Nematic Lyomesophase. *Mol. Cryst. Liq. Cryst.* **1983**, *103*, 243–253. [CrossRef]
37. Amaral, L.Q. Magnetic Orientation of Nematic Lyomesophases. *Mol. Cryst. Liq. Cryst.* **1983**, *100*, 85–91. [CrossRef]
38. Amaral, L.Q. The Lower Temperature Phase of a Nematic Lyomesophase System. *J. Appl. Crystallogr.* **1984**, *17*, 476–480. [CrossRef]
39. Amaral, L.Q. Transitions in a Lyomesophase: A Study by Thermal Analysis and Electron Microscopy. *Mol. Cryst. Liq. Cryst.* **1985**, *124*, 225–239. [CrossRef]
40. Amaral, L.Q.; Helene, M.E.M.; Bittencourt, D.R.; Itri, R. New nematic lyomesophase of sodium dodecyl sulfate. *J. Phys. Chem.* **1987**, *91*, 5949–5953. [CrossRef]
41. Amaral, L.Q.; Helene, M.E.M. Nematic domain in the SLS/H_2O/decanol system. *J. Phys. Chem.* **1988**, *92*, 6094–6098. [CrossRef]
42. Itri, R.; Amaral, L.Q. Study of the isotropic—Hexagonal transition in the system SLS/H_2O. *J. Phys. Chem.* **1990**, *94*, 2198–2202. [CrossRef]
43. Amaral, L.Q. First-Order Transition between Nematic Phases in Lyotropic Liquid Crystals. *Liq. Cryst.* **1990**, *7*, 877–881. [CrossRef]
44. Amaral, L.Q.; Gulik, A.; Itri, R.; Mariani, P. Micellar hexagonal phases in lyotropic liquid crystals. *Phys. Rev. A* **1992**, *46*, 3548–3550. [CrossRef]
45. Taddei, G.; Amaral, L.Q. Bending Energy and the relative stability of Micellar Forms. *J. Phys. Chem.* **1992**, *96*, 6102–6104. [CrossRef]
46. Amaral, L.Q. Micelles forming biaxial lyotropic nematic phases. *Liq. Cryst.* **2010**, *37*, 627–640. [CrossRef]
47. Itri, R.; Amaral, L.Q.; Mariani, P. Structure of the hexagonal phase of the sodium dodecyl sulfate and water system. *Phys. Rev. E* **1996**, *54*, 5211–5216. [CrossRef] [PubMed]
48. Spinozzi, F.; Carsughi, F.; Mariani, P.; Teixeira, C.V.; Amaral, L.Q. SAS from inhomogeneous particles with more than one domain of scattering density and arbitrary shape. *J. Appl. Crystallogr.* **2000**, *33*, 556–559. [CrossRef]
49. Lagerwall, J.P.F.; Scalia, G.; Haluska, M.; Dettlaff-Weglikowska, U.; Giesselmann, F.; Roth, S. Simultaneous alignment and dispersion of carbon nanotubes with lyotropic liquid crystals. *Phys. Stat. Sol. B* **2006**, *243*, 3046–3049. [CrossRef]
50. Lagerwall, J.P.F.; Scalia, G.; Haluska, M.; Dettlaff-Weglikowska, U.; Roth, S.; Giesselmann, F. Nanotube alignment using lyotropic liquid crystals. *Adv. Mater.* **2007**, *19*, 359–364. [CrossRef]
51. Thiele, T.; Berreta, J.-F.; Muller, S.; Schmidt, C. Rheology and nuclear magnetic resonance measurements under shear of sodium dodecyl sulfate/decanolwater nematics. *J. Rheol.* **2001**, *45*, 29–48. [CrossRef]
52. Forrest, B.J.; Reeves, L.W. New Lyotropic Liquid Crystals Composed of Finite Nonsphericai Micelles. *Chem. Rev.* **1981**, *81*, 1–14. [CrossRef]
53. Amaral, L.Q.; Santin Filho, O.; Taddei, G.; Vila-Romeu, N. Change in micelle form induced by cosurfactant addition in nematic lyotropic phases. *Langmuir* **1997**, *13*, 5016–5021. [CrossRef]
54. Henriques, E.F.; Henriques, V.B. Biaxial phases in polydisperse mean-field model solution of uniaxial micelles. *J. Chem. Phys.* **1997**, *107*, 8036–8040. [CrossRef]

55. Henriques, E.F.; Passos, C.B.; Henriques, V.B.; Amaral, L.Q. Mixture of changing uniaxial micellar forms in lyotropic biaxial nematics. *Liq. Cryst.* **2008**, *35*, 555–568. [CrossRef]
56. Mukherjee, P.K.; Sen, K. On a new topology in the phase diagram of biaxial nematic liquid crystals. *J. Chem. Phys.* **2009**, *130*, 141101. [CrossRef]
57. Do Carmo, E.; Liarte, D.B.; Salinas, S.R. Statistical models of mixtures with a biaxial nematic phase. *Phys. Rev. E* **2010**, *81*, 062701. [CrossRef] [PubMed]
58. Henriques, E.F.; Salinas, S.R. Biaxial nematic phase in the Maier-Saupe model for a mixture of discs and cylinders. *Eur. Phys. J. E* **2012**, *35*, 1–9. [CrossRef]
59. Henriques, E.F.; Henriques, V.B.; Krebs, P.R. Bimodal form distribution from modelling biaxial lyotropic liquid crystal solutions through a polydisperse Maier–Saupe model. *Liq. Cryst.* **2017**, *44*, 510–525. [CrossRef]
60. Mukherjee, P.K.; Rahman, M. Isotropic to biaxial nematic phase transition in an external magnetic field. *Chem. Phys.* **2013**, *423*, 178–181. [CrossRef]
61. Mukherjee, P.K. New phase diagrams in the mixture of rods and plates of biaxial nematic liquid crystals. *J. Mol. Liq.* **2016**, *220*, 742–746. [CrossRef]
62. Santos, O.R.; Braga, W.S.; Luders, D.D.; Sampaio, A.R.; Kimura, N.M.; Simões, M.; Palangana, A.J. Optical characterization of a biaxial nematic between uniaxial nematic lyotropic phases. *Phase Transit.* **2021**, *94*, 546–555. [CrossRef]
63. Chandrasekhar, S.; Ratna, B.R.; Sadashiva, B.K.; Raja, V.N. A Thermotropic Biaxial Nematic Liquid Crystal. *Mol. Cryst. Liq. Cryst.* **1988**, *165*, 123–130. [CrossRef]
64. Merkel, K.; Kocot, A.; Vij, J.K.; Korlacki, R.; Mehl, G.H.; Meyer, T. Thermotropic Biaxial Nematic Phase in Liquid Crystalline Organo-Siloxane Tetrapodes. *Phys. Rev. Lett.* **2004**, *93*, 237801. [CrossRef]
65. Müller, D.; Alexander, T.K.; Kierfeld, J. Chaining of hard disks in nematic needles: Particle-based simulation of colloidal interactions in liquid crystals. *Sci. Rep.* **2020**, *10*, 12718. [CrossRef] [PubMed]
66. Morange, M. *A History of Molecular Biology*; Cobb, M., Translator; Cambridge, Harvard University Press: Cambridge, UK, 1998.
67. Colotto, A.; Mariani, P.; Ponzi-Bossi, M.G.; Rustichelli, F.; Albertini, G.; Amaral, L.Q. Lipid-drug interaction: A structural analysis of pindolol effects on model membranes. *Biochim. Biophys. Acta BBA Biomembr.* **1992**, *1107*, 165–174. [CrossRef]
68. Amaral, L.Q.; Itri, R.; Mariani, P.; Micheletto, R. Structural study of the aggregates formed by the dinucleoside phosphate G2 in aqueous solution. *Liq. Cryst.* **1992**, *12*, 913–919. [CrossRef]
69. Mariani, P.; Amaral, L.Q.; Saturni, L.; Delacroix, H. Hexagonal-Cubic Phase Transitions in Lipid Containing Systems: Epitaxial Relationships and Cylinder Growth. *J. Phys. II* **1994**, *4*, 1393–1416. [CrossRef]
70. Mariani, P.; Amaral, L.Q. Micellar Growth in Hexagonal Phases of Lipid Systems. *Phys. Rev. E* **1994**, *50*, 1678–1681. [CrossRef] [PubMed]
71. Castelletto, V.; Itri, R.; Amaral, L.Q. Micellar aggregates near the isotropic-cubic liquid crystal phase transition. *J. Chem. Phys.* **1997**, *107*, 638. [CrossRef]
72. Castelletto, V.; Itri, R.; Amaral, L.Q.; Spada, G.P. Small-Angle X-Ray Scattering of DNA Fragments: Form and Interference Factors. *Macromolecules* **1995**, *28*, 8395–8400. [CrossRef]
73. Castelletto, V.; Amaral, L.Q. Short Range Order of Rodlike Polyelectrolytes in the Isotropic Phase. *Macromolecules* **1999**, *32*, 3469–3473. [CrossRef]
74. Castelletto, V.; Amaral, L.Q. Isotropic-Cubic Liquid Crystalline Phase Transition in Aqueous Solutions of PLPC. *J. Phys. Chem. B* **1999**, *103*, 8877–8881. [CrossRef]
75. Riske, K.A.; Amaral, L.Q.; Lamy-Freund, M.T. Thermal Transitions of DMPG bilayers in aqueous solution: SAXS structural studies. *Biochim. Biophys. Acta* **2001**, *1511*, 297–308. [CrossRef]
76. Riske, K.A.; Amaral, L.Q.; Döbereiner, H.-G.; Lamy, M.T. Mesoscopic structure in the chain melting regime of anionic phospholipid vesicles: DMPG. *Biophys. J.* **2004**, *86*, 3722–3733. [CrossRef]
77. Fernandez, R.M.; Riske, K.A.; Amaral, L.Q.; Itri, R.; Lamy, M.T. Influence of salt on the structure of DMPG studied by SAXS and optical microscopy. *Biochim. Biophys. Acta BBA Biomembr.* **2008**, *1778*, 907–916. [CrossRef] [PubMed]
78. Riske, K.A.; Amaral, L.Q.; Lamy, M.T. Extensive Bilayer Perforation Coupled with the Phase Transition Region of an Anionic Phospholipid. *Langmuir* **2009**, *25*, 10083–10091. [CrossRef]
79. Spinozzi, F.; Paccamiccio, L.; Mariani, P.; Amaral, L.Q. Melting Regime of the Anionic Phospholipid DMPG: New Lamellar Phase and Porous Bilayer Model. *Langmuir* **2010**, *26*, 6484–6493. [CrossRef]
80. Spinozzi, F.; Mariani, P.; Paccamiccio, L.; Amaral, L.Q. New lamellar phase with pores in the chain-melting regime of an anionic phospholipid dispersion. *J. Phys. Conf. Ser.* **2010**, *247*, 012019. [CrossRef]
81. Spinozzi, F.; Amaral, L.Q. Pore Model in the Melting Regime of a Lyotropic Biomembrane with an Anionic Phospholipid. *Langmuir* **2016**, *32*, 13556–13565. [CrossRef] [PubMed]
82. Rideau, E.; Dimova, R.; Schwille, P.; Wurm, F.R.; Landfester, K. Liposomes and polymersomes: A comparative review towards cell mimicking. *Chem. Soc. Rev.* **2018**, *47*, 8572–8610. [CrossRef]

83. Fenaroli, F.; Robertson, J.D.; Scarpa, E.; Gouveia, V.M.; Guglielmo, C.D.; De Pace, C.; Elks, P.M.; Poma, A.; Evangelopoulos, D.; Canseco, J.O.; et al. Polymersomes Eradicating Intracellular Bacteria. *ACS Nano* **2020**, *14*, 8287–8298. [CrossRef] [PubMed]
84. Rodríguez-Arco, L.; Poma, A.; Ruiz-Pérez, L.; Scarpa, E.; Ngamkham, K.; Battaglia, G. Molecular bionics-engineering biomaterials at the molecular level using biological principles. *Biomaterials* **2019**, *192*, 26–50. [CrossRef] [PubMed]
85. Amaral, L.Q. *Entre Sólidos e Líquidos: Uma Visão Contemporânea e Multidisciplinar Para O Ensino Medio*; Editora Livraria da Física: São Paulo, Brazil, 2014.
86. Amaral, L.Q. Mechanical analysis of infant carrying in hominoids. *Naturwissenschaften* **2008**, *95*, 281–292. [CrossRef]

molecules

MDPI

Article

Influence of Liquid Crystallinity and Mechanical Deformation on the Molecular Relaxations of an Auxetic Liquid Crystal Elastomer

Thomas Raistrick , Matthew Reynolds, Helen F. Gleeson and Johan Mattsson *

School of Physics and Astronomy, University of Leeds, Leeds LS2 9JT, UK; py14tr@leeds.ac.uk (T.R.); M.Reynolds@leeds.ac.uk (M.R.); h.f.gleeson@leeds.ac.uk (H.F.G.)

* Correspondence: k.j.l.mattsson@leeds.ac.uk

Abstract: Liquid Crystal Elastomers (LCEs) combine the anisotropic ordering of liquid crystals with the elastic properties of elastomers, providing unique physical properties, such as stimuli responsiveness and a recently discovered molecular auxetic response. Here, we determine how the molecular relaxation dynamics in an acrylate LCE are affected by its phase using broadband dielectric relaxation spectroscopy, calorimetry and rheology. Our LCE is an excellent model system since it exhibits a molecular auxetic response in its nematic state, and chemically identical nematic or isotropic samples can be prepared by cross-linking. We find that the glass transition temperatures (T_g) and dynamic fragilities are similar in both phases, and the T-dependence of the α relaxation shows a crossover at the same T^* for both phases. However, for $T > T^*$, the behavior becomes Arrhenius for the nematic LCE, but only more Arrhenius-like for the isotropic sample. We provide evidence that the latter behavior is related to the existence of pre-transitional nematic fluctuations in the isotropic LCE, which are locked in by polymerization. The role of applied strain on the relaxation dynamics and mechanical response of the LCE is investigated; this is particularly important since the molecular auxetic response is linked to a mechanical Fréedericksz transition that is not fully understood. We demonstrate that the complex Young's modulus and the α relaxation time remain relatively unchanged for small deformations, whereas for strains for which the auxetic response is achieved, significant increases are observed. We suggest that the observed molecular auxetic response is coupled to the strain-induced out-of-plane rotation of the mesogen units, in turn driven by the increasing constraints on polymer configurations, as reflected in increasing elastic moduli and α relaxation times; this is consistent with our recent results showing that the auxetic response coincides with the emergence of biaxial order.

Keywords: liquid crystalline elastomer; dielectric spectroscopy; shear rheology; polymer relaxations

1. Introduction

Liquid Crystal Elastomers (LCEs) are lightly cross-linked polymer networks with mesogen units incorporated within the main polymer chain, or as pendant units. LCEs combine the anisotropic behavior of liquid crystals, arising from the long-range orientational order of the mesogen units, with the rubber-like elastic behavior of conventional elastomers [1]. The macroscopic shape of LCEs is coupled to the ordering of the mesogen units making them stimuli response materials [2,3]. LCEs can also show a wide range of other useful properties such as stress–optical coupling [4], soft elastic deformation [5], biocompatibility [6] and enhanced damping properties [7].

To understand, and be able to predict material behavior, it is essential to understand the behavior of the relevant molecular or segmental relaxations. Examples include the links between relaxations and material aging and rejuvenation [8], the response to mechanical stress of polymer glasses [9–12], or the link between structural relaxation and ionic transport in polymer electrolytes [13]. Due to the unique properties of LCEs, including their often complex mechanical responses [5,7,14], it is important to characterize their relaxation dynamics. Furthermore, the relaxation dynamics and glass-formation in nematic liquid crystals in general is of significant fundamental interest [15–17], and LCEs allow for careful investigations of the behavior in nematic materials over a wide temperature range. The LCE in our study is particularly important in this regard, as it exhibits a molecular auxetic response [18,19] and chemically identical samples can be formed, by polymerization, in either the nematic or the isotropic phase. The existence of the nematic or isotropic phase for our LCE system has been demonstrated using Raman spectroscopy, where scalar order parameter ($S = 1/2\langle 3\cos^2\theta - 1\rangle$) values of 0.59 ± 0.05 and 0.00 ± 0.05 were found, respectively [4,19]. In addition, polarized optical microscopy and Berek compensator measurements confirm these results [4,20]. Moreover, differential scanning calorimetry (DSC) experiments show no evidence for any nematic-to-isotropic phase transition over the investigated temperature ranges, thus confirming that the phase is locked-in during synthesis [4,20]. Thus, the effects of the nematic order on glass formation and molecular relaxation can be directly, and elegantly, probed for a system of identical chemical composition. Furthermore, by determining the response on both rheology and relaxation dynamics to strain, offers important insight into the nature of the molecular auxetic behavior.

The structural (α) relaxation of a glass-forming liquid (or polymer melt), slows down dramatically upon cooling. If crystallization is avoided, e.g., by fast cooling, the material eventually falls out of equilibrium, resulting in a disordered solid—a glass. The glass transition temperature, T_g, is typically defined as the temperature (T) for which the characteristic α relaxation time, τ_α, reaches 100 s [21]. Upon cooling towards the glassy state, the molecular motions involved become increasingly cooperative [22,23], typically involving a few hundred molecules (or polymer segments) at T_g even though the detailed behavior is system-dependent [24]. In addition to the α-relaxation, glass-forming liquids or polymers typically show at least one additional secondary molecular relaxation that is generic to glass-formation and is linked to the α relaxation [25–27]; this is typically termed the β relaxation or the Johari–Goldstein β relaxation [28]. The β relaxation separates from the structural α relaxation below a temperature $\sim T_{\alpha\beta}$, persists in the glassy state, and since the glass structure is largely frozen (disregarding slow physical aging), its behavior can be characterized by a single fixed activation energy and is thus well-described by an Arrhenius expression:

$$\tau_\beta = \tau_0 exp\left(\frac{\Delta E_A}{k_B T}\right). \tag{1}$$

Here, τ_0 is a microscopic relaxation time ($\sim 10^{-13}$ s), ΔE_A is the activation energy and k_B is the Boltzmann constant. The α relaxation, on the other hand, typically shows a more complex non-Arrhenius temperature dependence with a T-dependent activation energy that grows for decreasing T. Empirically $\tau_\alpha(T)$ is often described using a so-called Vogel–Fulcher–Tammann (VFT) equation [21,22]:

$$\tau_\alpha = \tau_0 exp\left(\frac{DT_0}{T - T_0}\right), \tag{2}$$

where τ_0 is a microscopic relaxation time, T_0 is the temperature at which τ_α tends to infinity and D is a parameter which controls the extent of deviation of τ_α from Arrhenius behavior, the so-called 'fragility'; an alternative commonly used metric of fragility is $m = \mathrm{d}\log(\tau_\alpha)/\mathrm{d}(T_g/T)|T = T_g$. 'Fragile' liquids are highly sensitive to a T-change near T_g and are thus characterized by large m, or conversely, small D-values. In contrast, 'strong' liquids are characterized by small m, or large D-values [29]. Typically, strong liquids show near Arrhenius $\tau_\alpha(T)$ behaviour, whereas fragile liquids are highly non-Arrhenius [21,23,29]. An empirical VFT function can often describe the $\tau_\alpha(T)$ behavior well over an extended T-range above T_g. However, for temperature above $T \sim T_B$, where $T_B \sim 1.2 - 1.6 \times T_g$ (the T_B/T_g ratio is system-dependent and has shown a systematic variation with fragility [30]), the T-dependence often crosses over into another VFT-like behavior, with a more significant non-Arrhenius T-dependence [31]. This dynamic crossover, at T_B, generally coincides relatively well with $T_{\alpha,\beta}$ [25,27] and with several other changes in the liquid including a decoupling of translational and rotational diffusion [21] and a change in the T-dependence of the α-relaxation strength [22]. Furthermore, the relaxation behavior for $T > T_B$ is often relatively well-described by so-called ideal-Mode Coupling Theory (ideal MCT) [22], but the correspondence with its predictions break down for $T \sim T_B$. Thus, T_B signifies a fundamental change in the liquid dynamics, which is an observation predicted already in the 1960s by Goldstein [32] and associated with changes in molecular relaxations due to the need to overcome energy barriers within the experienced 'energy landscape' that become significantly larger than $k_B T$ for $T < T_B$. Finally, at high T, above a temperature T_A ($T_A > T_B > T_g$), the need for cooperative motions in the α relaxation disappears, or is significantly reduced, resulting in Arrhenius behavior with a fixed energy barrier [31].

The general glass-transition phenomenology for polymeric glass-formers is similar to that of non-polymeric systems. However, due to chain connectivity and the corresponding presence of intra-molecular degrees of freedom, the inter-relation between the α relaxation and the secondary relaxations is more complicated [33,34]. Moreover, for sufficiently long oligomers (longer than a Kuhn or Rouse bead) or polymers, the difference between the $\tau_\alpha(T)$ behaviour on either side of T_B is often reduced, or disappear altogether [24,35], meaning that only one effective VFT is observed across a wide T-range; the origin of this behavior is not presently well understood. Additionally, for polymers, the transition to Arrhenius behavior at high-T is often difficult to study due to polymer degradation [23]. Cross-linked polymer systems, in turn, show similar relaxation behavior to other polymeric glass-forming materials. However, the presence and density of cross-links can affect the relaxation behavior, and the effects on the structural α-relaxation are typically to slow it down, resulting in increased T_g [36].

For LCEs, there have been relatively few studies focusing on molecular relaxation behavior [37–39]. The nomenclature for glass-transition-related relaxation dynamics of LCEs typically follows that used for side-chain liquid crystal polymers (SCLCPs). In SCLCPs, 4 relaxation processes are typically observed: δ, α, β and γ, named in order of increasing relaxation frequency for a fixed T [22]. The structural (α) relaxation in SCLCPs involves the backbone polymer segments and is directly related to the glass transition [22,40]. The δ process has been observed in both SCLCPs [22,41–43] and LCEs [37–39] and is typically slower than the α relaxation, follows a VFT dependence, and is attributed to reorientation of the mesogenic units around the polymer backbone. The faster secondary relaxations, typically termed β and γ are generally assigned to motions of the mesogenic units, where the β relaxation is typically assigned to fluctuations of the mesogen around its molecular long axis [22,41,44,45], and the γ relaxation is assigned to motions of either the spacer unit, or the terminal group of the side-chain mesogen [40,41].

In SCLCPs, just as for non-LC glass formers, a cross-over in dynamic behavior has been observed for $\tau_\alpha(T)$. However, in contrast to the non-LC behavior, for SCLCPs the low-T VFT behavior typically changes to a higher T Arrhenius behavior at a crossover temperature, $T^* \sim 1.1 - 1.3 \times T_g$ [42,46]. Interestingly, T^*/T_g is similar to T_B/T_g, as observed for non-LC systems, suggesting similarities in their origins. Moreover, this cross-over to Arrhenius behavior is typically only observed for polymers that form LC phases. For example, in one study, the existence of LC phase behavior was removed from a SCLCP by the substitution of a hydrogen on the biphenyl mesogen group with the bulkier methoxy (-OCH3) group. This resulted in a loss of the crossover from VFT to Arrhenius at T^*, and τ_α could instead be described by a single VFT [42]. This observation suggests that fluctuations related to LC phases are required for this behavior to occur.

It is important to understand how LCEs respond to applied deformations. LCEs are typically characterized by two different types of response to an applied strain. The first is the Semi-Soft Elastic (SSE) response [5,47]. Here, the elastic cost of deformation is reduced by the continuous rotation of the nematic director in counter-rotating domains known as 'stripe domains' [47]. In the semi-soft elastic response, it is assumed that the nematic order of the system remains constant whilst the orientation of the director is free to rotate in response to the applied strain [1,47]. Thus, for SSE LCEs, it has been shown that the shear storage modulus perpendicular to the director is smaller than parallel to the director, $G'_{\parallel} > G'_{\perp}$ [48,49]; this 'softening' is believed to be due to the rotation of the director. A number of investigations have been performed to determine the rheological behavior and also the relaxation behavior for typical LCEs that undergo SSE. Examples include LCEs based on siloxane [37,38,48–50] or acrylate [39] chemistry, with a side-chain mesogenic attachment. The second class of LCEs deforms via a completely different mechanism known as the mechanical Fréedericksz transition [19,20,51]. Here, a discontinuous rotation of the director is observed upon application of strain, as opposed to the continuous rotation of SSE. Before the discontinuous rotation occurs, the director is essentially fixed and only the degree of nematic order changes with applied strain [1,52]. Additionally, in some of the LCEs that deform via a mechanical Fréedericksz transition, a *negative* Poisson ratio in one of the transverse axes is observed [18], and this behavior has been identified as a molecular auxetic response [18,19]. LCEs that deform via the mechanical Fréedericksz transition, which include the LCE of this study, typically have an acrylate backbone [20,51]. The underlying mechanism that causes an LCE to deform via a discontinuous rotation instead of displaying the SSE response is still to be determined, and the theoretical framework behind the two mechanisms appear to be different [1]. To address this and better understand the LCEs of the second class, we here perform a detailed characterization of both the molecular relaxations and rheological behavior for an acrylate-based LCE which is previously known to deform via the mechanical Fréedericksz transition. Additionally, we investigate how both applied strain and the liquid crystal phase affects the molecular relaxation dynamics and the rheological response of an all-acrylate LCE.

2. Results and Discussion

2.1. Relaxation Dynamics of the Isotropic and Nematic LCE

Broadband Dielectric Spectroscopy (BDS) was performed for both an unstrained nematic and isotropic LCE sample, and the complex dielectric permittivity $\varepsilon^*(f) = \varepsilon'(f) - i\varepsilon''(f)$ was determined over a wide frequency range (~10^{-2}–10^6 Hz) (for details see the Section 3). Results for the dielectric loss, $\varepsilon''(f)$, of the nematic LCE are shown in Figure 1 for a few selected temperatures; the dielectric response over the full temperature range is included in the Supplementary Materials (SM) (Figure S1). For both LCE samples, three relaxation processes are identified, where α denotes the structural relaxation, directly related to the glass-transition, and β and γ, corresponding to more local motions, are characterized by shorter timescales. In addition, a clear contribution from ionic DC-conductivity is present, as evidenced by the power law contribution $\varepsilon'' \propto f^{-1}$ observed at low frequencies in Figure 1a.

LCEs previously investigated in literature also typically show a so-called δ relaxation that is slower than the α relaxation and associated with the motions of the mesogenic unit around its short axis [37–39]. To investigate whether there is evidence for any dielectrically active molecular relaxation slower than the α relaxation, in the dynamic range where DC-conductivity dominates the dielectric response, we use two additional approaches: (i) We estimate the dielectric loss free of conductivity effects ε''_{der} by conversion from the ε' spectra using a well-established simple approximation of the Kramers–Kronig transformation [53]:

$$\varepsilon''_{der}(\omega) = -\frac{\pi}{2}\frac{\partial\varepsilon'(\omega)}{\partial\ln\omega} \text{ where } \omega = 2\pi f\,. \quad (3)$$

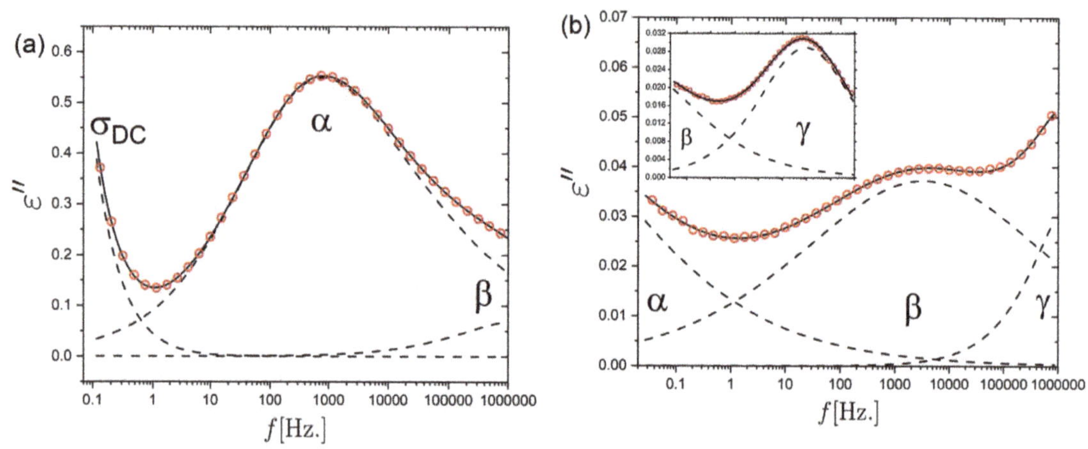

Figure 1. Dielectric loss versus frequency for the nematic LCE sample. The fits to the data, described in the main text, are shown in solid lines, and the individual contributions from the α, β and γ relaxations, as well as the DC conductivity (σ_{DC}) are labelled and shown in dashed lines. Data for different temperatures are shown in: (**a**) $T = 315.15$ K, (**b**) main graph: $T = 248.15$ K and inset: 163.15 K.

A pure DC-conductivity is only observed in the dielectric loss, ε'', and not in ε'; thus, this approach can significantly reduce the interference from DC-conductivity in the analysis. (ii) We analyzed data in the dielectric modulus representation $M^* = 1/\varepsilon^*$, for which the DC-conductivity contribution is generally suppressed [53]. For both these approaches, a slow Debye-like relaxation was identified in the dynamic window slower than the α relaxation (see Supplementary Materials). This Debye-relaxation coincides with the presence of electrode polarization, as observed by a low-frequency increase in ε'. We thus interpret this relaxation peak as arising from so-called conductivity relaxation [43,53], due to electrode polarization, and thus not to a molecular relaxation; electrode polarization is due to charge accumulation at the sample-electrode interface [22]. Given the interference from DC-conductivity and electrode polarization at low frequencies, we cannot completely rule out that a slow δ process exists. However, we do not find any evidence for it within the investigated temperature and frequency range. LCEs, previously studied in the literature, that display a δ relaxation, also show a transition from an isotropic phase to an LC phase. Conversely, the LCE investigated here shows no such transition [4,20], nor is there evidence for the δ relaxation, as discussed above; This suggests that despite the presence of side-chain mesogenic units, the larger-scale motions of the mesogenic units are hindered in both the isotropic and nematic phases of the LCE. The cross-link density of our LCE, based on chemical composition, is 7.1 mol%. This cross-linking density is comparable to the 7.5 mol% cross-link density in a previous study of an LCE with a comparable T_g value, which did show a δ relaxation [38], suggesting that the level of crosslinking in our LCE might not be enough to trap the mesogen large-scale movements. Thus, we instead suggest that the presence of the non-mesogen pendant units (EHA; see MS) in our system, could entrap the mesogen A6OCB sidechain, in turn preventing the larger-scale motions corresponding to the $\delta-$ relaxation.

To investigate the T-dependent complex permittivity $\varepsilon^*(f)$, the data are fit by a sum of relaxation contributions and a contribution for the DC-conductivity (see Section 3). The α relaxation is fit with a Havriliak–Negami (HN) expression:

$$\varepsilon^*(f) = \varepsilon_\infty + \frac{\Delta\varepsilon}{(1 + (i2\pi f \tau_{HN})^p)^q} \quad (4)$$

where ε_∞ is the high-frequency permittivity, $\Delta\varepsilon$ is the dielectric strength, τ_{HN} is the HN characteristic timescale, p and q are shape parameters of the response function; p corresponds to the low-frequency power law exponent and $p \times q$, the high-frequency exponent of the relaxation. The α relaxation was found to have similar broadness and asymmetry (based on p and $p \times q$ values) in the isotropic and nematic LCE. The T range for which the α relaxation is in the probed frequency window is ~290 K to 360 K. As an example, at 319 K the HN parameters characterizing the α relaxation are $p \approx 0.5$ and $p \times q \approx 0.2$ for both the isotropic and nematic LCE. The secondary β and γ relaxations were well-described by the simpler Cole–Cole expression, for which q was set to 1 in Equation (4); this leads to a relaxation peak that is symmetric on a logarithmic frequency axis. The β relaxation was found to have a fairly constant value of p, with mean values 0.21 ± 0.01 and 0.25 ± 0.01 for the isotropic and nematic phases, respectively. The p value for the γ relaxation was found to increase in a linear fashion from ~0.25 to ~0.4 for increasing temperature. We choose the most probable relaxation time, corresponding to the peak maximum, as the characteristic timescale for each relaxation contribution; The peak relaxation times are defined as $\tau_p = (2\pi f_p)^{-1}$ where τ_p is the frequency corresponding to the peak maximum. For the HN function, which is generally asymmetric (i.e., $q \neq 0$), the HN timescale is not the timescale of the peak maximum, but τ_p can be derived from τ_{HN}, p and q (see Section 3). The T-dependent characteristic relaxation times for the α, β and γ relaxations are shown in an Arrhenius plot in Figure 2. The α relaxations for the isotropic (Iso) and nematic (Nem) LCE samples are here fitted using a VFT expression, whereas the β and γ relaxations are fitted using an Arrhenius expression. The fitting parameters resulting from the fits are provided in Table 1.

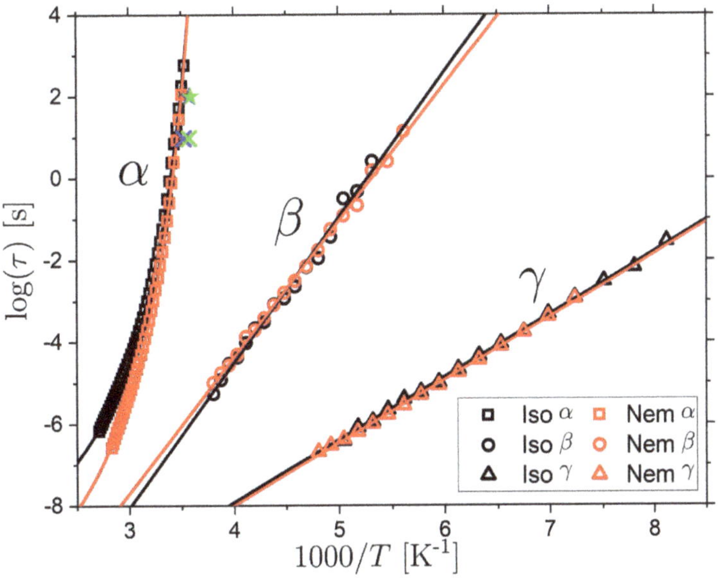

Figure 2. Characteristic relaxation times for the α (squares), β (circles) and γ (triangles) relaxations identified for the isotropic (black) and nematic (red) LCE samples. The results of a VFT fit to the α relaxation data, and Arrhenius fits the β and γ relaxation data are shown in solid lines. Data for the glass transition temperature (α relaxations) determined from modulated DSC (stars) are also shown as corresponding to a time-scale of $\tau_0 = 9.56$ s, for the isotropic (blue) and nematic (green) LCE sample. Glass-transition temperature data using DSC performed at a fixed rate of 10 °C/min (crosses) are plotted assuming a corresponding time-scale of τ_α = 100 s (see text for further discussion).

Table 1. Results of Arrhenius and VFT fits to relaxations in the isotropic and LCE nematic samples.

Sample	Process	τ_0 (s)	ΔE_A (kJ mol^{-1})	D	T_0 (K)	T_g (K)
Isotropic	α	5.3×10^{-11}	-	5.0	243	286
	β	2.0×10^{-19}	67.9	-	-	-
	γ	9.3×10^{-15}	29.3	-	-	-
Nematic	α	1.5×10^{-12}	-	5.6	243	285
	β	2.1×10^{-18}	63.6	-	-	-
	γ	7.2×10^{-15}	29.4	-	-	-

As seen from the fits in Figure 2 and the corresponding parameters, the β and γ relaxations have very similar T-dependencies in the isotropic and nematic states. The corresponding activation energies ΔE_A for the β relaxation are 67.9 kJ mol^{-1} and 63.6 kJ mol^{-1} for the isotropic and nematic LCE, respectively. The corresponding ΔE_A results for the γ relaxation are 29.3 kJ mol^{-1} and 29.4 kJ mol^{-1}, respectively. The results of the Arrhenius fits thus demonstrate that the LCE phase has little effect on the β and γ relaxations. This may be due to the relatively short characteristic length scales of these relaxations, in comparison to the relevant length scale of the LC phase. It is interesting to compare our LCE results to those of other LC systems in the literature. A collection of activations energies, ΔE_A, for β relaxations in polyacrylate and polymethacrylate SCLCPs are found in work by Kremer and Schönhals [22] and Schönhals and Hans-eckartcarius [40], ΔE_A values within the range 46.5–68.9 kJ mol^{-1} were reported. The exact ΔE_A value depends on the terminal group attachment of the mesogenic unit and the length of the alkyl spacer between the backbone and the mesogenic unit. Our acrylate-based LCE has a spacer length of 6. The activation energy of the SCLCP with the closest chemistry to our LCE (acrylate backbone, spacer length of 6) is 62.8 kJ mol^{-1}, which is close to our ΔE_A values of 67.9 kJ mol^{-1} and 63.6 kJ mol^{-1} for the isotropic and nematic LCE respectively. Thus, we follow the literature assignment and associate the observed β relaxation with fluctuations of the mesogen around its long axis. Literature values of the γ relaxation, typically assigned to motions of the alkyl spacer units, have ΔE_A values in the range $\sim 33 - 35$ kJ mol^{-1} [41,54,55], which is close to the ΔE_A of ~29 kJ mol^{-1} determined for the γ relaxation in the isotropic and nematic LCE. Hence, based on comparison to literature data, the γ process for our LCE is likely due to motions of the 6-alkyl chain connecting A6OCB to the acrylate backbone.

The characteristic timescale of the α-relaxation, $\tau_\alpha\ (T)$, follows a non-Arrhenius T-dependence in both the isotropic and nematic phase, and we use an empirical VFT expression to describe the behavior (Table 1). However, as discussed in the introduction, for molecular glass formers the T-dependence of the α relaxation can often not be described accurately using a single VFT equation. Thus, to further investigate the detailed T-dependence of $\tau_\alpha(T)$ we perform a derivative analysis of the data, as first suggested by Stickel et al. [31,56]. By plotting the parameter Z vs. $1000/T$ where Z is given by:

$$Z = \left(\frac{d \log \tau_\alpha}{d\left(\frac{1000}{T}\right)} \right)^{-\frac{1}{2}}, \tag{5}$$

a VFT-behavior is linearized and this analysis has been shown to be useful in identifying changes in the T-dependence [31,56].

As shown in Figure 3a, both the isotropic and nematic LCE samples undergo a change in $\tau_\alpha(T)$ at a temperature $T^* \approx 333$ K, corresponding to $1000/T^* \approx 3.0$. For $T < T^*$, the gradients of the linear fits are similar (-0.36 and -0.33 for the isotropic and nematic samples, respectively). However, for $T > T^*$, the gradients are clearly significantly different (-0.21 and -0.04). A clear change in $\tau_\alpha(T)$ for both the isotropic and nematic samples are therefore supported by the derivative analysis, as evidenced by a change in gradient at T^*. Moreover, it is clear that the gradient for the nematic samples for $T > T^*$ is very near zero and can thus be well described by an Arrhenius behavior. To further investigate $\tau_\alpha(T)$, the data are fit with separate VFTs, for $T < T^*$ (solid lines) and $T > T^*$ (dashed lines), as shown in Figure 3b. For the nematic data, for $T > T^*$, the data were also fit using an Arrhenius expression for comparison. The results of the fits are outlined in Table 2.

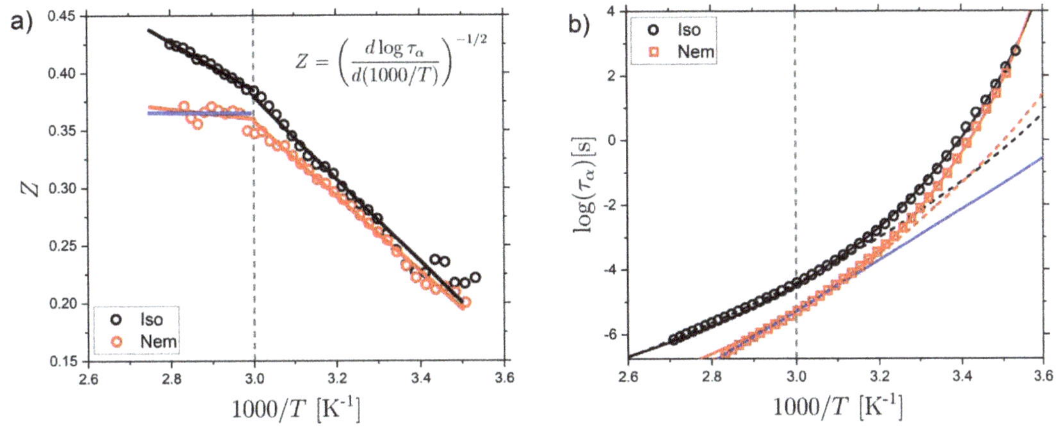

Figure 3. (**a**) Derivative analysis of $\tau_\alpha(T)$ data (see text for details) for the isotropic (black) and nematic (red) LCE samples, respectively. A crossover behavior is observed at $T^* \approx 333$ K, as shown with a dashed line. Linear fits of the data are shown both for temperatures above and below T^*. For the nematic samples the data for $T > T^*$ can be well described using a horizontal line (corresponding to Arrhenius behavior). (**b**) Arrhenius plot of $\tau_\alpha(T)$ vs. inverse temperature for the isotropic and nematic samples. For $T < T^*$, the VFT fits to the data are shown in solid lines. For $T > T^*$ VFT fits are shown in dashed lines for the isotropic (black) and nematic (red) sample. An Arrhenius fit is also applied to the nematic data before $1000/T = 3.0$ (blue solid line).

From the fit result of the dielectric data, the T_g values can be determined from $T_g = T(\tau_\alpha = 100 \text{ s})$, which results in T_g values of 286 K and 285 K, respectively. These T_g values can be compared with those obtained from m-DSC and DSC experiments. (see Supplementary Materials for DSC traces) The DSC measurements, performed on cooling at 10 K/min, give T_g values of 279 K and 278 K, which is relatively close to T_g from BDS. The modulation period of m-DSC (60 s) probes the α relaxation on a timescale of ~10 s (see Section 3 for full details). The values of T_g from m-DSC are 283 K and 279 K for isotropic and nematic LCE. Thus, the trend where $\tau_\alpha(T)[\text{iso}] > \tau_\alpha(T)[\text{nem}]$ for $T > T_g$ is confirmed by the m-DSC measurements. Furthermore, the similar VFT parameters $D = 5.4$ and $D = 5.1$ reflect that $\tau_\alpha(T)$ behave in a very similar manner for temperatures approaching T_g and that the fragility of the two samples is thus similar. The fragility parameter (m) for the isotropic and nematic LCEs can also be determined from the VFT parameters, and the results are $m = 110$ and $m = 130$ for the isotropic and nematic LCE, respectively. This demonstrates that both LCE phases are fragile glass-formers with fragility values consistent with those of polymers [23,30]. For $T > T^*$, however, both the isotropic and nematic LCEs are less fragile which is demonstrated by the VFT parameters $D = 16.6$ and $D = 17.2$. The derivative analysis demonstrates that the nematic LCE can be well described as Arrhenius in this T-range. The gradient of the Stickel plot for $T > T^*$ is -0.04 for the nematic LCE which is very close to Arrhenius behavior (gradient of 0). Thus, an Arrhenius fit is also applied to this region of the data of the nematic LCE (Figure 3b, solid blue line). The result of the Arrhenius fit is 1.15×10^{-29} s and 151 kJ mol^{-1}. Interestingly, following the same trend towards more Arrhenius-like behavior above T^*, the isotropic LCE is less non-Arrhenius above, than below, T^*.

Table 2. Results of the VFT and Arrhenius fits for the isotropic and nematic LCE.

Sample	1000/T	τ_0 (s)	ΔE_a (kJ mol^{-1})	D	T_0 (K)	T_g (K)
Isotropic	< 3.0	5.0×10^{-14}	-	16.6	183	270
	> 3.0	2.5×10^{-11}	-	5.4	241	286
Nematic	< 3.0	2.7×10^{-16}	-	17.2	193	275
	< 3.0	1.15×10^{-29}	151.0	-	-	-
	> 3.0	3.8×10^{-12}	-	5.1	245	285

Finally, we find that, T^* is situated well above T_g ($T^*/T_g \approx 1.17$) and corresponds to $\tau_\alpha(T^*) = 3.8 \times 10^{-5}$ s and 5.0×10^{-6} s for the isotropic and nematic LCE samples, respectively. Importantly, the ratio T^*/T_g for the cross-over in behavior in the LCEs is close to the ratio T_B/T_g observed for conventional glass formers (T_B/T_g = 1.2–1.6) [30], suggesting a related origin. However, for conventional non-LC glass formers, $\tau_\alpha(T)$ typically transitions to more markedly non-Arrhenius (more fragile) behavior for $T > T_B$, whereas we observe the opposite trend for $T > T^*$. Moreover, for non-LC glass-formers, a bifurcation scenario resulting in a Johari–Goldstein β relaxation at $T_{\alpha\beta} \sim T_B$ is often observed. Neither of the two dielectrically active secondary relaxations, β or γ, observed in this work show any relation with T^*. We note, however, that we cannot rule out the existence of another secondary relaxation that is not dielectrically active (and thus not detected in our measurement), which demonstrates a bifurcation behavior near T^*.

In SCLCPs, a qualitatively similar $\tau_\alpha(T)$ behavior is observed for the α relaxation, where an Arrhenius behavior describes the data for $T > T^*$, and a VFT behavior for $T < T^*$ [42,46,57]. Temperature ratios of T^*/T_g = 1.1–1.3 have been observed in nematic and smectic SCLCPs [42,46], which is close to the ratios observed for our LCE. In studies of homologous series of methacrylate-based SCLCPs with systematically varying side-chain length, the T^*/T_g ratio was found to be fixed, independent of the phase transition temperatures, thus supporting the independence of the crossover behaviour on the LC phase behavior [42,46]. The activation energy, ΔE_A, for the α relaxation of SCLCPs in the high-T Arrhenius regime typically ranges from 80–127 kJ mol^{-1} [42] and can be compared to the value of 151 kJ mol^{-1} obtained in our work. Thus, the behaviors observed for SCLCPs are generally very similar to the observations for our LCE.

The cross-over behavior observed for SCLCPs has often been suggested to be related to a matching between the characteristic length-scale of correlated motions involved in the structural α relaxation, (often discussed in terms of a cooperatively rearranging region, or CRR) and a length-scale characterizing microphase separation of mesogen-rich and polymer-rich domains [22,42,46]. Microphase separation has been observed in polysiloxane SCLCPs that readily phase separate [58] and in SCLCPs which form layers due to smectic phase behavior. However, we observe cross-over behavior at T^* in both the nematic LCE and isotropic LCE; microphase separation is certainly not present in our isotropic LCE. Thus, at least for our LCE, the origin of the observed crossover behavior lies elsewhere, and we will return to this in Section 2.3.

2.2. *Ionic Conductivity Behavior of the Isotropic and Nematic LCE*

There is strong interest in developing polymer-based materials for applications in energy materials, such as batteries, e.g., as electrolytes or electrode binders [13,59]. Polymer-based electrolytes could, e.g., provide both the safety, mechanical flexibility and rigidity needed for ion-transporting membranes to act simultaneously both as ion conductors and electrode separators. Elastomers, in particular, show promise since their cross-links impart mechanical stability while T_g can be kept relatively low, which provides mobility and thus more efficient ion transport. However, polymer-based materials still have relatively high T_g values, which means that if ion transport is strongly coupled to the structural relaxation, sufficient ion transport is very difficult to achieve [60,61]. Thus, it is of significant interest for future applications to understand how to control the coupling of ion transport and structural relaxation. Moreover, polymer systems with LC functionalities have been identified as interesting candidates for battery applications [62,63] due to the additional structural control provided, which can affect both the nature and efficiency of the ion transport, as well as allow for anisotropic control of charge transfer. LCEs are particularly interesting in this respect due to their cross-linked nature which results in a combination of mechanical rigidity and liquid crystalline functionalities. However, there have been few studies to date exploring this for LCEs [64,65].

Studies of conventional polymeric materials have shown that for relatively low-T_g polyethers, such as PEO and PPG, the structural α relaxation and the ionic DC-conductivity (σ) are highly coupled whereas higher T_g, less flexible polymers such as polycarbonate and poly(methyl methacrylate) show significant decoupling [60,66,67]. It is noteworthy that the well-coupled polyethers have strongly ion-coordinating ether oxygens spaced regularly along the backbone and ion transport has been shown to preferentially take place along the chain for these [67,68]; thus, these commonly used ion-conducting polymers might be regarded as outliers compared to polymers without such clear coordination structures. The detailed origin of the 'decoupling behavior' is however not presently well understood. For inorganic superionic glasses which demonstrate very strong decoupling, models exist which are typically focused on understanding the contributions to the fixed energy barrier that controls ion motion in the glass, e.g., from electrostatic and elastic forces [61]. Similar approaches could be adapted also for polymeric materials, where T-dependent changes in elastic and dielectric properties for $T > T_g$ would result in a T-dependent barrier, as typically described using a VFT expression [61], and a corresponding degree of decoupling.

The decoupling parameter, γ, is a way to quantify the extent of the decoupling of the ionic conductivity from the α relaxation and is determined from the relationship $\sigma \propto \tau_\alpha^{-\gamma}$. For long-chain polymers, a variation of the decoupling parameter between different polymers have been observed, and a rough trend was suggested where polymers with higher fragility [60] (typically also higher T_g) showed more decoupling than less fragile polymers. Dynamic fragility can typically be related to molecular packing and as a rule of thumb it was thus proposed that more fragile polymers pack less effectively, and thus leave more space for ions to move without assistance from matrix relaxations [60,66].

In our study, we have not specifically added ions. However, our LCE samples contain a small number of ionic impurities and we determined the corresponding ionic DC-conductivity and investigated how this correlates with the structural relaxation, and importantly identify the effects of the LC phase on this behavior. To determine the decoupling parameter, γ, we plot the measured ionic conductivity vs. the inverse structural α relaxation time for both the isotropic and nematic LCEs in a double-logarithmic representation, as shown in Figure 4. This type of Walden-like plot [59] is often used to investigate the relationship between ionic DC-conductivity and structural relaxation, and a slope near 1 indicates a strong coupling, whereas a smaller γ indicates decoupling. We find that the isotropic LCE is relatively well coupled with a coupling parameter of $\gamma = 0.87$, whereas the nematic LCE is significantly less coupled, corresponding to a $\gamma = 0.54$. The fragility of the isotropic and nematic LCE is $m = 110$ and $m = 130$, respectively. These two fragility values are quite similar, but we still note that a higher fragility is typically associated with a stronger decoupling in conventional polymeric materials [60].

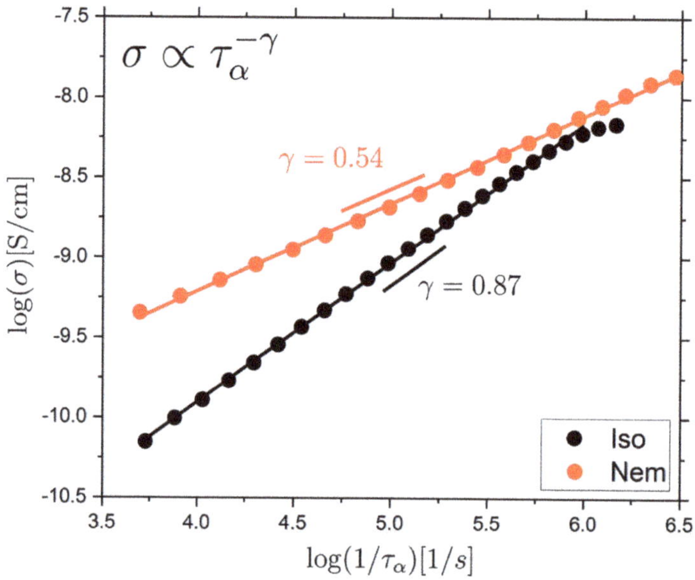

Figure 4. Plot of $\log(\sigma)$ vs. $\log(1/\tau_\alpha)$ for the isotropic and nematic LCE samples. Furthermore shown is the γ coupling coefficient from the equation $\sigma \propto \tau_\alpha^{-\gamma}$, which is related to the extent of coupling between the α relaxation and the ionic conductivity.

For non-polymeric glass-formers, the Stokes–Einstein (SE) relation relates the translational diffusion coefficient D to the viscosity η, according to $D = k_B T / a\eta r$, where r is the radius of the diffusing entity, a is a constant, and T is the temperature. The SE relation often holds both for probe diffusion and molecular self-diffusion for $T > T_B$; thus, $D \propto \eta^{-1}$. Moreover, since $\eta \sim \tau_\alpha$ [69] to a good approximation, $D \propto \tau_\alpha{}^{-1}$. However, for $T < T_B$, a more complex 'fractional' SE behavior is often observed instead, where $D \propto \tau_\alpha{}^{-\zeta}$, with $\zeta \sim 0.6 - 0.9$ [70,71]; ζ has been reported to vary with fragility [72]. Since the ionic DC-conductivity is proportional to the diffusion coefficients for the ions in the material, the similarity between these observations and the observations for ion-conducting glass-forming materials, described above, is evident.

A commonly invoked explanation for the 'breakdown' of the SE-relation is that it is caused by the development of dynamic heterogeneities for $T < T_B$, i.e., different spatial regions in the liquid are characterized by different characteristic relaxation times. The detailed link is not presently clear, but it has been argued that the SE-breakdown occurs since D and η (or τ_α) are averaged differently over the heterogeneous distribution of environments [73,74]. Alternatively, it has been suggested that D and η (or τ_α) couple differently to spatial variations in intermolecular cooperativity or that the relationship is affected by the presence of an emerging secondary relaxation mechanism [75]. In summary, the detailed picture is presently not clear and more work is clearly needed to determine these links.

For LC-based materials, a number of studies have investigated the coupling between ion conduction and structural relaxation [15,17,76]. In 5CB, in the isotropic phase on the approach of T_{NI}, a strong deviation from a Walden-plot gradient of 1 was observed, indicating significant decoupling, which was explained as due to the presence of pre-transitional nematic fluctuations [16]. The addition of nanoparticles to the isotropic phase of 5CB was shown to reduce the decoupling between DC-conductivity and the α relaxation and the behavior was attributed to the nanoparticle-induced disruption of pre-transitional nematic fluctuations in the isotropic phase [76]. The nematic phase shows a distribution of nematic domains with slightly varying order parameters and orientations throughout the sample, and is therefore likely to be more dynamically heterogeneous than the isotropic phase, which contains only pre-transitional nematic effects. Thus, explanations focused on the presence and strength of dynamic heterogeneities, and how these influence both the structural relaxation and ionic conductivity, might explain the observations of a greater decoupling in the nematic than in the isotropic phase observed in our LCE system and LC systems in general.

Finally, it is important to note that for the nematic LCE, the mesogenic units are arranged in an ordered manner which could in itself affect the transport of ions. Effects on ionic transport, and the relationship between the DC-conductivity and the structural relaxation, due to induced spatial anisotropies have also been observed for the non-LC polymer PEO, where the PEO chains were aligned either by mechanical stretching or by magnetic and electric fields [68,77]. The observed effects were interpreted as due to molecular structure-induced changes in conduction pathways. For LC-based systems, the structural organization characterizing some phases, could thus directly affect both the efficiency of the ion transport, as well as the coupling between the ion transport and the structural relaxation. What is clear from our study is that LCEs containing relevant ion-coordinating chemistries should be highly interesting materials for which the LC phases can be utilized to tune the ion transport properties.

2.3. *Rheology of the Isotropic LCE Sample*

The rheological behavior of LCEs is fundamentally interesting due to the coupling of the mesogenic units to the polymeric network. Whilst there have been previous studies on the dynamic rheological behavior of LCEs these have largely been performed on LCEs with polysiloxane backbones and/or LCEs which deform via the SSE response [48–50,78]. Here, we determined the rheological behavior of our isotropic LCE, which in contrast to these literature studies has an acrylate-based backbone. We studied the LCE using Dynamic Mechanical Analysis (DMA) and Small Amplitude Oscillatory Shear rheology (SAOS). To obtain data over a wide frequency range, we used Time Temperature Superposition (TTS) to construct master curves (see the Section 3 for a detailed description of the procedure). The validity of TTS was first investigated by plotting the data in a so-called van Gurp–Palmen representation (see Supplementary Materials) which removes all explicit time-dependence from the unshifted rheological data and therefore, shows if accurate TTS using frequency shifts is possible [79]. Both the DMA and SAOS data fall on a single line on the van Gurp–Palmen plot, respectively. Thus TTS can be adequately performed on these samples with $T = 40\ ^\circ\mathrm{C}$ selected as the reference temperature. A horizontal shift factor is applied to the data at other temperatures to form a master curve (see Materials and Methods for details). The resulting rheological master curves are shown in Figure 5, SAOS is used to determine the complex shear modulus, $G^*(\omega) = G'(\omega) + iG''(\omega)$, ($G'$ = green circles, G'' = blue circles) and DMA is used to determine the complex Young's modulus $E^*(\omega) = E'(\omega) + iE''(\omega)$ (E' = hollow black circles, E'' = hollow red circles). The DMA data are shifted (shifted E' = black circles, shifted E'' = red circles) to directly compare the SAOS and DMA data to each other. As seen in Figure 5, E' and E'' can be collapsed onto G' and G'' (vertical shift of -0.37 applied). The Poisson ratio for the isotropic LCE can be determined from these data using the relationship:

$$G' = \frac{E'}{2(1+\nu)}, \tag{6}$$

where G' and E' are the shear and elastic storage moduli respectively and ν is the Poisson's ratio. The Poisson's ratio is determined in the region where we have data for both E' and G', i.e., between $\sim$0.6 and 1.9×10^2 rad/s. The mean value of the Poisson ratio over the full range of the rheological data is 0.19 ± 0.03. The value of the Poisson's ratio at 1 rad/s is 0.25 ± 0.05, where this data has been selected for the reference temperature and therefore removes any added complications resulting from TTS shifting. These determined values of the Poisson's ratio (0.19 ± 0.03 and 0.25 ± 0.05 respectively) fall within the physical limits of the Poisson's ratio for isotropic materials ($-1 \leq \nu \leq 0.5$), and are comparable to those determined in previous studies for polydomain LCEs, where $0.2 < \nu < 0.35$ was observed for small strains, depending on the cross-link density [80,81].

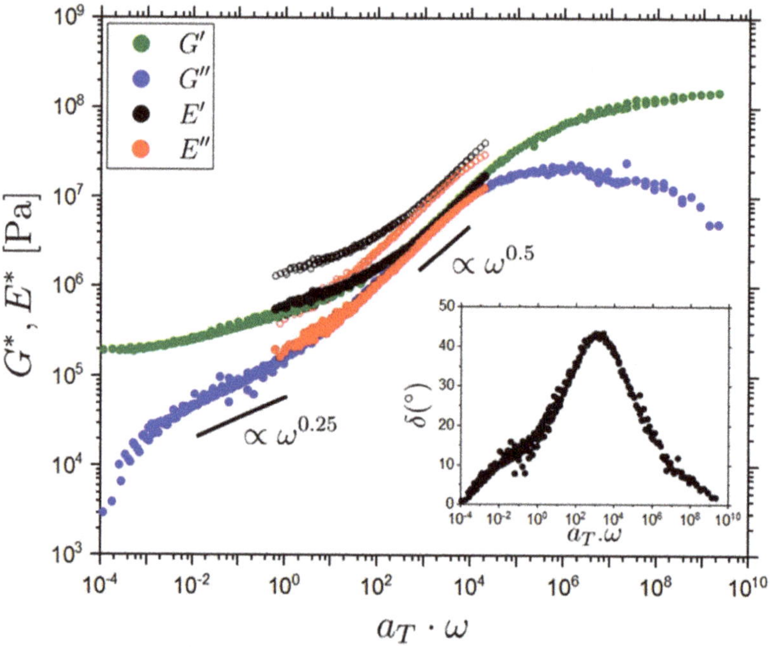

Figure 5. TTS master curves for shear storage modulus (G', green circles), shear loss modulus (G'', blue circles), unshifted storage modulus (E', hollow black circles), unshifted loss modulus (E'', hollow red circles), shifted storage modulus (E', black circles) and shifted loss modulus (E'', red circles). A vertical shift of -0.37 (on the log-scale) is applied between G^* and E^* demonstrating a very good agreement between the two data sets. Approximate power-law scalings of $\omega^{0.5}$ and $\omega^{0.25}$ discussed in the text are illustrated. Inset: phase angle ($\delta = \tan^{-1}(G''/G')$) against TTS shifted angular frequency ($a_T.\omega$) for the SAOS data.

The rheological data in Figure 5. includes the structural α-relaxation response in the high-frequency range ($\sim 10^7 - 10^{10}$ rad/s). At lower frequencies in the range of $\sim 10^2 - 10^4$ rad/s, a power law-like regime is observed where $G' = G'' \propto \omega^{0.5}$, and this scaling is more pronounced for G'' due to the transition towards a rubber-like plateau at low frequencies in G', resulting from the presence of permanent cross-links. The observed scaling is evidence of a Rouse-like spectrum [82,83]. A Rouse-like mode spectrum has similarly been reported in isotropic, nematic and smectic LCE systems [48–50,78]. Towards lower frequencies ($10^1 - 10^{-2}$ rad/s), a flatter approximate power-law-like regime of $G'' \propto \omega^{0.25}$ is observed. This contribution is also observed as a shoulder in the low-frequency flank of the peak in the loss angle (δ) shown in Figure 5 (inset, $\tan\delta = G''/G'$). Similarly, a low-frequency power law of $G' \approx G'' \propto \omega^{0.3}$ has been observed in the SmA phase of LCE systems, whereas it was not observed for the corresponding isotropic phase [50,78]. This scaling in SmA LCEs was interpreted as due to the presence of smectic layers which influence the otherwise separated polymer backbones [50,78,84]. However, for our isotropic LCE this situation is clearly not the same, and the observed behavior must have a different origin. Our LCE is a randomly cross-linked network containing the pendant units A6OCB and EHA. Based on. this, we propose two mechanisms for the low-frequency behavior. Firstly, the observed response could be related to the motion of free chains through the network. A $G'' \propto \omega^{0.2-0.3}$ dependence has indeed been reported in cross-linked poly(dimethylsiloxane) (PDMS) networks where linear 'free' PDMS chains were present [85]. Secondly, the relaxation could be due to the motions of the dangling pendant chains within the network. This has been observed for poly(butyl acrylate) networks, where it has been interpreted as due to chain motions linked to pendant arm retraction, [86], and in PDMS networks with pendant chains where the details of the loss contribution depended on the pendant chain length [87].

From the TTS shift-factors, we directly obtain information about the T-dependent characteristic time-scale for the LCE within the investigated T-range. The α relaxation timescale at the reference T (40 °C) is determined from the peak maximum in G''. Subsequently, $\tau_\alpha(T)$, from the rheological data, is determined by applying the $a_T(\omega)$ to $\tau_\alpha(T_0 = 40\ ^\circ\text{C})$ (see Section 3 for details). The temperature dependence of the α relaxation determined from rheology is compared to our results for BDS in Figure 6. Here, a vertical shift of 2.58 (on the log scale) is applied to the rheology data set to overlay it with the BDS data. A shift of the relaxation time scales is expected between rheological and dielectric measurements [88] and based on this analysis, good agreement is found between the two data sets. Next, the rheology data are fit with a VFT expression and the results of the fits are shown in Table 3. Importantly, the close correspondence between the two data sets supports the validity of the TTS approach used in our analysis.

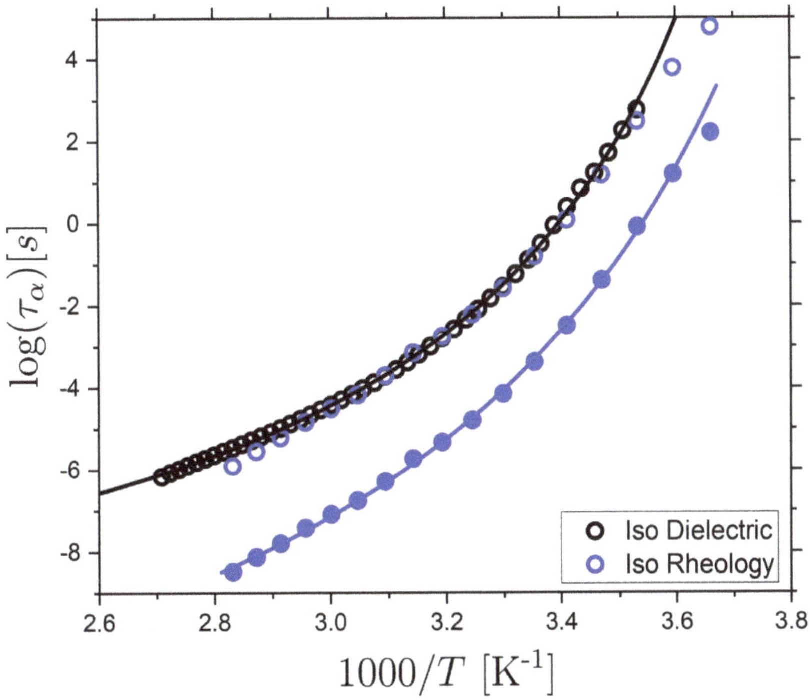

Figure 6. Arrhenius plot showing the α relaxation time, $\tau_\alpha(T)$ vs. inverse T for the isotropic LCE as measured using BDS (black hollow circles) and SAOS (filled blue circles). The SAOS data are also shown with a vertical shift of 2.58 (blue open circles), demonstrating the similarity in the T-dependence of the α relaxation, as probed by the two techniques. VFT fits of the BDS and SAOS data are shown in solid lines.

Table 3. Results for the VFT-fits of the SAOS data for the isotropic LCE sample.

τ_0 (s)	D	T_0 (K)	T_g (K)
1.33×10^{-15}	8.3	227	276

2.4. *Volume of Correlated Motions in the Isotropic LCE*

As the glass transition is approached, the α relaxation is characterized by the presence of dynamic heterogeneities, i.e., regions in space with dynamics different from their surroundings. The size of the spatial regions of correlated motion has been determined for both non-polymeric and polymeric liquids [89], using a range of techniques, including NMR [90], modulated DSC [91], dielectric spectroscopy [89], or MD-simulations [92]. The determined length-scale characterizing correlated motions is typically ~1–5 nm [91,93] in the vicinity of T_g, where the detailed behavior depends on the specifics of the system, such as its dynamic fragility. Here, we use our BDS data to estimate the characteristic length scale (see the Section 3 for a detailed description). The dynamic heterogeneity is linked to fluctuations of the dynamics in time and space and these can be characterized by a so-called 4-point dynamic susceptibility $\chi_4(t)$ that quantifies the amplitude of spontaneous fluctuations around the average dynamics [94]. $\chi_4(t)$ can be expressed as a correlation, in time and space, of 2-point correlators that are readily experimentally accessible, e.g., through BDS. $\chi_4(t)$ is typically a non-monotonic function with a peak occurring near the α-relaxation time, and a height that is proportional to the volume of correlated motions, $V_{corr,\,4}$, or alternatively the number of molecular units that undergo correlated motions within this volume, $N_{corr,4}$ [94]. It is difficult to experimentally directly measure spontaneous fluctuations and thus $\chi_4(t)$, however, using a fluctuation-dissipation relation, it has been demonstrated that one can approximate $\chi_4(t)$ by determining induced fluctuations, e.g., how the relevant 2-point correlator (experimentally readily available) responds to a perturbation, such as temperature [89,94].

Here, we use this technique, as outlined in detail in the Section 3, to determine an estimate of $V_{corr,4}$. The T-dependent correlation volume $V_{corr,4}$ estimated from the approach outlined above is shown in Figure 7 as a function of inverse temperature for the isotropic LCE; in the Supplementary Materials $V_{corr,4}$ is also plotted versus τ_α. The observed behaviour is typical for glass-formers with an increasing $V_{corr,4}$ for decreasing T and the stronger T-dependence at higher temperatures, which significantly reduces near T_g [89,95]. We do note that the most significant change in T-dependence of $V_{corr,4}$ takes place at around 350 K ($1000/T \sim 2.86$), which is above the range where we observe changes in $\tau_\alpha(T)$ at $T^* = 333$ K. Importantly, we find an estimate for the volume of correlated motions for the isotropic LCE at the transition temperature T^* of $\sim$1 nm^3, corresponding to a length-scale of $l_\alpha(T^*) \sim 1$ nm.

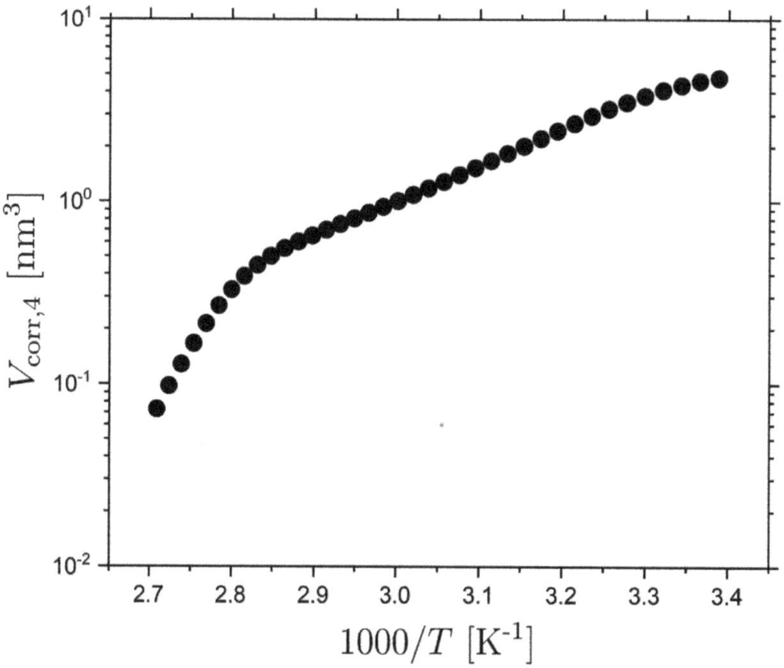

Figure 7. Volume of correlated molecular motions for the isotropic LCE sample as a function of inverse temperature, as discussed in detail in the text.

Given that $\tau_\alpha(T)$ undergoes a crossover to more Arrhenius-like behavior for $T > T^*$ for both the nematic and isotropic LCE, which is opposite to the crossover to more non-Arrhenius behavior for non-LC glass-formers, it is interesting to compare $l_\alpha(T^*)$ to any length-scale related to nematic behavior for the 'isotropic' LCE. The most pertinent length-scale is the static correlation-length ξ of pre-transitional nematic domains [96]. The correlation length of the pre-transitional nematic regions (ξ) in an isotropic phase follow the equation [96]:

$$\xi = \xi_0 \sqrt{\frac{T_c}{T - T_c}}, \tag{7}$$

where ξ_0 is the bare correlation length of the pre-transitional nematic regions with $\xi_0 \sim 0.5$ nm in simple molecular LCs [96], T_c is a supercritical temperature which is typically ~1 K lower than T_{NI}. The isotropic phase is templated into the LCE and after polymerization there is no evidence of a phase change in the isotropic LCE as investigated by DSC [4]. We expect that the pre-transitional nematic regions are frozen-in during polymerization, a phenomenon similar to 'frozen-in order' near cross-linking points and 'quenched disorder' previously observed in LCEs [97,98]. Our isotropic LCE is polymerized at 60 °C and the T_{NI} of the precursor mixture is 36 °C [4,20]; therefore by substituting these values into the above equation, we expect that the correlation length will be of the order of 3.5 ξ_0. Using the typical value of $\xi_0 = 0.5$ nm, we thus expect the correlation length of the nematic domains in the isotropic LCE to be of the order of ξ = 1.8 nm. This is comparable to the length scale of correlated motions in the α relaxation at T^*. Thus, it seems plausible that the difference in $\tau_\alpha(T)$ in the isotropic and nematic phase is related to the interplay between length-scales characterizing the correlated molecular (segmental) motions of the α relaxation and the correlation length of the pre-transitional phenomena in the form of nematic fluctuations of the isotropic LCE.

The results of this analysis are in general agreement with studies of the structural relaxation in molecular LCs [15–17,99]. It has been shown that the temperature-dependent behavior of the structural relaxation of LC materials is strongly influenced by pre-transitional phenomena in the isotropic phase [15]. It has also been shown that the addition of nanoparticles disrupts the pre-transitional phenomena present in the isotropic phase [76,100]. With low concentrations of nanoparticles, a cross-over from non-Arrhenius to Arrhenius behavior has been observed in the isotropic phase; further increase in the nanoparticle concentration causes a cross-over from Arrhenius to non-Arrhenius behavior [100]. Hence the difference in $\tau_\alpha(T)$ in the isotropic and nematic LCE could be understood by the presence of pre-transitional nematic regions and nematic correlations, respectively, and the relative size of these with respect to the length-scale of correlated motions of the α relaxation. Previous measurements on SCLCPs in their isotropic phase, thus lacking pre-transitional nematic fluctuations, have shown that $\tau_\alpha(T)$ can be described with a single VFT [42], as is also typical for polymeric materials with a sufficiently long chain-length [24]. In contrast, the precursor mixture presented herein has a nematic to isotropic transition and thus, in the isotropic phase, will have pre-transitional nematic regions present.

2.5. *Effect of Strain on the Dielectric and Rheological Behaviors of the Nematic Sample*

In previous work on our LCE system [18,19], it has been demonstrated that the nematic phase shows a complex response to an imposed uniaxial deformation; this includes linear and non-linear elastic behavior, a reduction in uniaxial order, the emergence of biaxial order, and molecular auxetic response [18,19]. The stress–strain behavior has been reported previously to reveal an initial linear elastic regime, followed by a plateau-like softening behavior, and a subsequent growth of the stress [14, 20]. The plateau and subsequent stiffening of the stress–strain curve have been observed in both the engineering stress–engineering strain and true stress–engineering strain and are therefore not a consequence of sample necking [14]. A softening of the stress–strain response is a hallmark of the 'semi-soft elastic response' [5,47], which is due to the continuous rotation of the nematic director in counter-rotating domains known as stripe domains. However, it has been previously demonstrated that the behavior observed for our nematic LCE is instead due to a mechanism known as the mechanical Frèedericksz transition, which is signified by a discontinuous rotation of the nematic director [19,20]. Investigations into the order parameter of our nematic LCE have shown a link between the mechanical Frèedericksz transition, reduction in uniaxial order and the emergence of biaxial order with imposed strains [19]. To understand these observations, it is important to also identify, and understand, the effects of deformation on the rheology and relaxation dynamics of the nematic LCE. To achieve the former, the complex Young's modulus was determined using a Dynamic Mechanical Analyzer (DMA) (see Section 3 for details). Measurements were performed on a sample with dimensions 5 cm × 0.2 cm × 100 μm. The LCE film was subjected to varying elongations, and thus strains, ranging from 0 to 120%. The elongations of the LCE sample is shown in true strain (ϵ_t) representation:

$$\epsilon_t = \ln\left(\frac{L_f}{L_i}\right), \tag{8}$$

where L_f is the length of the sample after elongation, and L_i is the initial sample length. After the applied elongation, the LCE is left to stress–relax for 2 min, where this time is selected to be sufficiently short to avoid sample breakage, yet long enough to not affect the auxetic response of the material [18]. Similar relaxation times have been used to investigate the tensile mechanical response of the material and the order parameter behavior [14,19]. For each elongation, its relevant storage and loss elastic modulus ($E'(f_0)$ and $E''(f_0)$) determined for $f_0 = 1$ Hz ($\omega \approx 6.3$ rad/s) were determined by applying oscillatory strains of 0.1% to the pre-elongated sample; oscillatory strains of 0.1% were confirmed to be in the linear viscoelastic region of the nematic LCE via a strain sweep. For comparison, tensile stress–strain measurements, published previously [20], are shown in the inset from which the Young's modulus (E) was determined. The results and methodology of the tensile measurements have been reported in full elsewhere [18], but briefly, tensile stress–strain measurements were performed in a bespoke rig consisting of two actuators and a load cell enclosed in a temperature-controlled environment; images of the LCE are recorded with a camera and changes in length and width of the sample are determined which allows for the calculation of the true-stress and true-strain of the LCE. The trues tress is defined as $\sigma_t = F/A$, where F is the force measured on a sample and A is the cross-sectional area of the sample after the application of strain. The cross-sectional area of the LCE as a function of strain is calculated using the assumption of constant volume, which has previously been shown to be a good approximation for this LCE [18].

The effects of elongational strain on the mechanical response (e.g., the Young's or shear moduli) is an important consideration in understanding materials under strain [101,102]. Changes in loss moduli could, e.g., be related to the breaking of bonds or molecular slippage occurring within polymers [101]. Other examples include experiments on natural rubber and styrene butadiene rubber for which strains < 170% did not affect the complex moduli (for $f = 1$ Hz), whereas at larger strains an increase was observed in both storage and loss moduli [103] which was assigned to an increase in the effective constraints on the molecular orientations within the elastomer network due to the applied strain [104], and to the finite extensibility of the network [105].

Figure 8 shows the evolution of E' and E'' for the nematic LCE as a function of external strain applied perpendicular to the nematic director. The grey dashed line in Figure 8 denotes the onset of the molecular auxetic response that has been reported for this system; this response is related to out-of-plane rotations of the mesogenic units [18,19]. The black and red dashed lines are exponential growth functions which serve as guides to the eye. At low values of applied strain ($\epsilon_t < 0.22$), $E'(f_0)$ and $E''(f_0)$ are relatively constant with average values of 4.9 ± 0.2 MPa and 2.7 ± 0.1 MPa respectively; the elastic modulus determined from the gradient of the true stress–true strain data in this region ($\epsilon_t < 0.22$) is 4.6 MPa showing an excellent agreement with our determined $E'(f_0)$. Between $\epsilon_t = 0.22$ and $\epsilon_t = 0.53$ average values of E' = 7.7 ± 0.6 MPa and E'' = 4.7 ± 0.4 MPa are observed as compared to 4.9 ± 0.2 MPa and 2.7 ± 0.1 MPa for $\epsilon_t < 0.22$; therefore in this regime ($\epsilon_t < 0.55$) there is a small dependence of dynamic moduli on strain. Within error, $\tan(\delta) = E''/E'$ is constant ($\tan(\delta) = 0.55 \pm 0.03$ and $\tan(\delta) = 0.60 \pm 0.07$ for $\epsilon_t < 0.22$ and $0.22 < \epsilon_t < 0.55$ respectively) which suggest that the LCE is deforming elastically (albeit non-linearly) throughout this strain region.

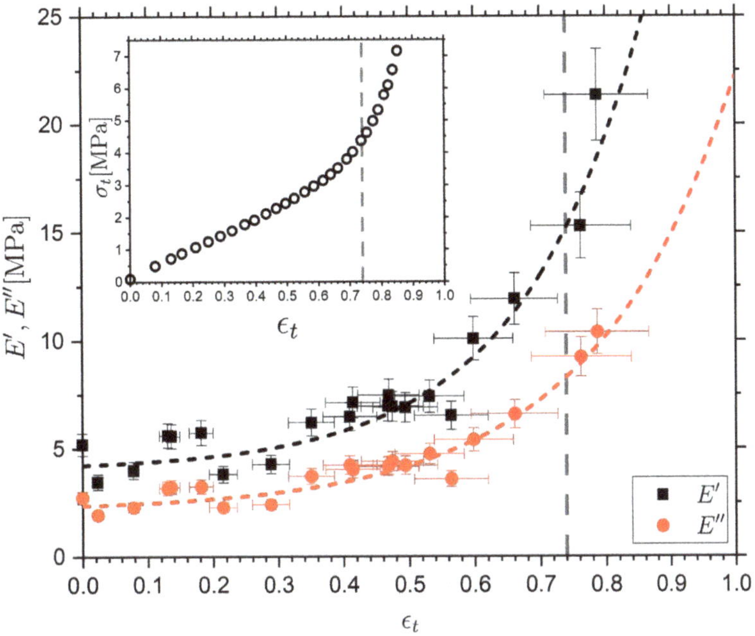

Figure 8. Storage and loss moduli determined at 1 Hz (2π rad/s) and T = 21 °C, as a function of true strain applied perpendicular to the nematic director. Inset: non-dynamic tensile tests showing the true stress (hollow circles) (σ_t [MPa]) as function of true strain ϵ_t. The grey dashed line marks the threshold for the onset of the molecular auxetic response.

At larger values of strain ($\epsilon_t > 0.55$) a stiffening of the stress–strain response is observed in the true stress–true strain representation, additionally a significant increase in $E'(f_0)$ and $E''(f_0)$ is observed. At the strain corresponding to sample breakage ($\epsilon_t \approx 0.78$), $E'(f_0)$ = 21.3 MPa which is consistent with the value of the elastic modulus E~24 MPa determined from the gradient of the true stress–true strain data in this region. All the way up to sample breakage, tan(δ) is constant, within error, and the sample thus behaves elastically throughout the whole operational strain range. We note that the upturn in the true stress–true strain response, and the corresponding increases in $E'(f_0)$ and $E''(f_0)$ occurs just before the sharp rotation of the director at the mechanical Fréedericksz transition, and the emergence of molecular auxetic response (signified by the grey vertical line) [18]. To summarize, the DMA measurements reveal that the LCE sample deforms elastically throughout the whole strain range, and the observed upturns in $E'(f_0)$ and $E''(f_0)$ occur near, but before, the onset of the mechanical Fréedericksz transition.

To further understand the effects of applied strain on the nematic LCE, the α relaxation response for different applied strains was investigated using BDS. Literature reports of BDS experiments on a lightly cross-linked (6.5% mol/m^3) polyisoprene-based non-LC elastomer found no effects of an applied "static" strain on the α-relaxation [9]. However, a polyurethane-based elastomer containing 32.5 wt.% hard segments (4,4′-diphenylmethane diisocyanate and 1,4-butanediol) subjected to applied strains ranging from 0% to 300%, showed a broadening and slowing-down of the α relaxation, and an increase in the fragility parameter, m [9]. A similar increase in the fragility, and slowing down of the α relaxation has also been reported for a polyurea-based elastomer [106]. In both cases, the change in the dynamics of the strained elastomers was related to increasing constraints imposed on the soft segments of the elastomers due to a deformation-induced reduction in microphase separation [9,106]. Experiments have also been performed aimed at investigating the effects of mechanical deformations on the relaxation dynamics of polymer glasses [10–12,107]. Uniaxial deformation experiments performed on poly(methyl methacrylate) (PMMA) under either constant load or constant strain-rate demonstrate that below the yield stress, the mobility of the α relaxation is enhanced by stress which can be interpreted as due to tilting of the potential energy landscape, leading to a lowering of activation barriers [108]. Above yielding, more dramatic behavior can be observed with a strong sensitivity to applied stress or strain-rate, and significant effects on the observed spatial dynamic heterogeneity [12,107].

To perform BDS measurements under strain, the nematic LCE sample was stretched to the desired strain and affixed to a 20 mm brass plate with Kapton tape. The sample dimensions were nominally 7 cm × 1.5 cm × 100 μm (L × W × T) and a brass plate of 5 mm was placed on top of the sample to allow for BDS measurements. It was confirmed that the Kapton tape was placed sufficiently far away from the electrodes to not influence the measurements. After mounting, the sample was left to stress–relax for 2 min before a measurement was performed. All measurements were performed at T = 23 °C to be able to directly compare with the DMA measurements described above. Figure 9 shows the frequency-dependent dielectric loss $\varepsilon''(f)$ data, as normalized by the maximum of the relaxation peak corresponding to the α relaxation, $\varepsilon''_{\mathrm{p}}$. Data are shown for a set of applied strains varying over the range 0 to 140% ($\epsilon_t = 0$ to 0.88). Within the measured dynamic range ($10^{-2} < f < 10^6$ Hz), the α relaxation is observed together with the low-frequency side of the β relaxation. The data are thus fit using a sum of an HN-contribution (α relaxation) and a CC-function (β relaxation). We here focus on the α relaxation results, since the β relaxation contribution is only partly covered in the dynamic window. It is clear from Figure 9a that the general shape of the α relaxation remains the same even at large values of applied strain. The characteristic α relaxation times τ_α (corresponding to the maxima of the loss peaks) obtained for increasing values of applied strain are shown in Figure 9b. Two separate LCE samples were investigated to determine the effect of strain on the α relaxation, one taken to smaller strain values ($\epsilon_t \leq 0.44$) (open triangles), and one taken to larger strain values ($\epsilon_t \leq 0.86$) (open circles) which is above the characteristic strain for the mechanical Fréedericksz transition where the LCE displays a molecular auxetic response ($\epsilon_t = 0.73$).

Due to slight differences in the unstrained α relaxation timescale between the two samples, the $\tau_{\alpha(\epsilon_t)}$ data are normalized by the unstrained timescale ($\tau_{\alpha(\epsilon_t=0)}$), the result of which is shown in Figure 9b. The dielectric spectra and unnormalized relaxation timescales are shown in the Supplementary Materials. For the sample subjected only to lower values of strain (Figure 9b, open triangles), we find no obvious trend in the τ_α timescales within the accuracy of the data. However, in the sample taken to higher true strains (Figure 9b, open circles) there is a clear shift in the α relaxation to slower relaxation times. The results are therefore consistent with findings of a stress-relaxed polyurethane elastomer containing rigid units [9], which were attributed to increased strain-induced constraints on the polymeric backbone.

To summarize, both $E'(f_0)$ and $E''(f_0)$ remain relatively unchanged with increasing strain until a large strain (~0.55) is imposed, after which an increase in both moduli is observed. Correspondingly, τ_α remains relatively unchanged until an applied strain of ~0.55 after which the relaxation slows down. Thus, the changes in τ_α and the complex modulus are observed for comparable values of true strain. (Figure 9b) Both of these effects could be understood in terms of a strain-induced increase of constraints, e.g., reflected in a reduced configurational entropy of the polymer backbone [9,109], and the finite extensibility of the network [105]. The increases in E', E'' and τ_α occur near the discontinuous rotation of the director (the mechanical Frèedericksz transition) and the onset of the molecular auxetic response [18–20]. We cannot exclude the possibility that the observed behavior is a coincidence. However, we recently suggested that the auxetic response in our LCE is related to the emergence of biaxial order linked to out-of-plane rotations of the mesogenic units [19]. We propose that the out-out-plane rotations occur due to the strain-imposed configurational restrictions on the polymer backbone.

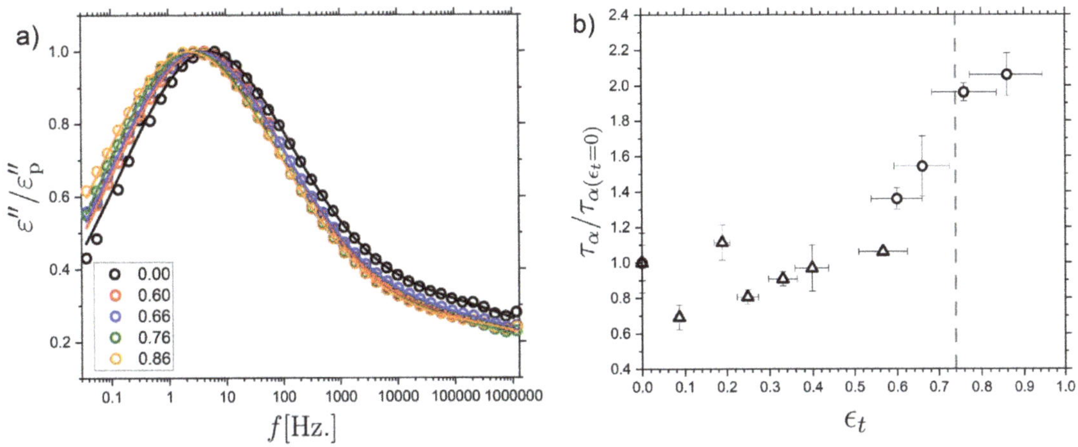

Figure 9. (**a**) Normalised dielectric loss ε'' versus frequency at T = 23 °C for nematic LCE sample, measured for a range of different applied true strains ϵ_t, from 0.00 to 0.86 as shown in the legend. The solid lines are fit to the data, described in detail in the text. (**b**) The α-relaxation times τ_α versus applied true strain, as determined from fits to the data shown in panel (**a**), and described in detail in the text. The grey dashed line marks the threshold for the onset of the molecular auxetic response.

3. Materials and Methods

3.1. Synthesis of the Liquid Crystalline Elastomer

The liquid crystalline elastomer used in this study was synthesized following a protocol published previously [4,18,20], and we thus only briefly describe the synthesis here. A mixture of the reactive mesogens 6-(4-Cyano-biphenyl-4′-yloxy)hexyl acrylate (A6OCB), 1,4-bis-[4-(6-acryloyloxyhexyloxy)benzoyloxy]-2-methylbenzene (RM82) and the non-reactive mesogen 4′-Hexyloxybiphenyl (6OCB) is melted at 100 °C and subsequently cooled to 40 °C, and the non-mesogenic spacer 2-ethylhexyl acrylate (EHA) and photopolymeriser methyl benzoylformate (MBF) are added to form an isotropic mixture of the precursor chemicals. The molar ratios of this mixture are outlined in Table 4 and the corresponding chemical structures of the components are shown in Figure 10. The isotropic monomer mixture is capillary-filled at 40 °C into a cell with a spin-coated poly vinyl alcohol (MW = > 89,000, 0.5 %wt) layer that is rubbed to provide alignment. To form a monodomain nematic sample the sample is cooled to room temperature, into the nematic phase, and allowed to align for 20 min. The sample is polymerized using a UV curer (2.5 mW cm^{-2}) for 2 h. An isotropic sample is prepared by capillary filling into an unaligned cell and curing at 60 °C. Both samples are washed with a 70/30% methanol/dichloromethane mixture to remove the unpolymerized 6OCB, and finally dried for 4 h at 60 °C. The prepared samples have nominal dimensions of 7 cm × 1.5 cm × 100 μm.

Table 4. Chemical names and mol% for the precursor mixtures of the isotropic and nematic LCEs.

Chemical Name	Mol% of Monomer Mixture
A6OCB	14.6
RM82	7.1
6OCB	55.9
EHA	20.9
MBF	1.5

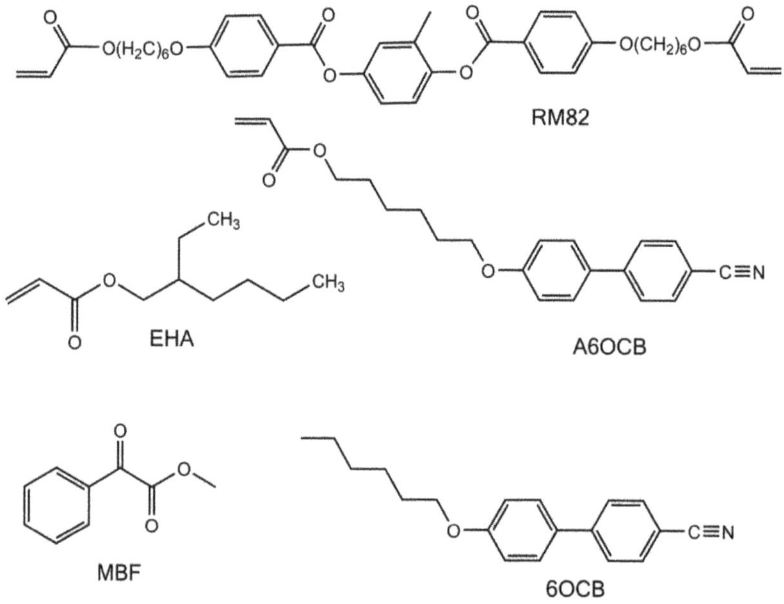

Figure 10. Structures of the constituent chemicals for both the nematic and isotropic LCEs.

3.2. *Broadband Dielectric Spectroscopy*

Broadband dielectric spectroscopy was performed over a frequency range of $\sim 10^{-2} < f < 10^6$ Hz. Using a Novocontrol Alpha-A dielectric analyser The LCE samples were sandwiched between two 10 mm diameter circular metal electrodes, separated by 100 μm using silica spacers. The temperature was controlled using a Novocontrol Quatro Cryosystem with an accuracy of 0.1 K and experiments were performed between −150 °C and 100 °C. The temperature-dependent complex permittivity was analyzed using a sum of relaxation contributions and a contribution from DC-conductivity. Each relaxation contribution was described using a Havriliak–Negami (HN) or Cole–Cole (CC) expressions [22,110], and the full relaxation spectrum is thus described by:

$$\varepsilon^*(f) = \varepsilon_\infty + \sum_j^N \frac{\Delta\varepsilon_j}{\left(1 + \left(i2\pi f \tau_{HN,j}\right)^{p,j}\right)^{q,j}} + \frac{-i\sigma}{2\varepsilon_0 \pi f}, \tag{9}$$

where $\varepsilon^*(\omega)$ is the complex permittivity, ε_∞ is the high-frequency limit of the complex permittivity, N is the total number of relaxation processes, and the index j refers to a particular relaxation. $\Delta\varepsilon_j$ is the dielectric strength, and $\tau_{HN,j}$ is a characteristic time-scale of relaxation j, and p, j and q, j are parameters characterizing the stretching of the jth relaxation process, respectively. σ is the DC-conductivity due to the presence of ionic impurities. The α relaxation was fitted using the HN function ($p \neq 1$, $q \neq 1$) whilst the β and γ relaxations are fitted using the CC expression ($p \neq 1, q = 1$). We consistently use the maximum of the dielectric loss, corresponding to the most probably relaxation time, as the relaxation time characterising a particular relaxation. To obtain the time-scale corresponding to the peak maximum, τ_p , we use the expression [22]:

$$\frac{1}{\tau_p} = \frac{1}{\tau_{HN}} \left(\sin\left(\frac{p\pi}{2+2q}\right)^{1/p} \sin\left(\frac{pq\pi}{2+2q}\right)^{-1/p} \right). \tag{10}$$

3.3. *Rheology*

Small amplitude frequency-dependent rheology was performed both in tension and shear. A Dynamic Mechanical Analyzer (DMA) (Rheometrics Solid Analyser; RSAII) with a film tension attachment was used to determine the frequency-dependent complex Young's modulus $E^*(\omega) = E'(\omega) + iE''(\omega)$, where E' is the storage and E'' the loss modulus. A strain of 0.1% ensured operation within the linear viscoelastic (LVE) regime, as confirmed by strain sweeps prior to the frequency scans. Frequency scans were subsequently performed between 6.3 and 78.5 rad/s from T = 22 °C to 48 °C in 2 °C steps. Moreover, Small Angle Oscillatory Shear (SAOS) measurements were performed on a Rheometrics ARES strain-controlled rheometer, using a liquid nitrogen cooling system in combination with a forced-convection oven. Samples were loaded between 5 mm and diameter parallel plates, using a gap of 0.5–1.0 mm. A strain-sweep test was carried out to ensure that measurements were performed in the linear regime, and frequency-scans were performed between 0.1 and 100 rad/s, every 5 degrees over the temperature range of 80 to 0 °C to encompass the α relaxation response. Upon lowering the temperature, the gap was reduced to ensure that the sample remained in the correct shape, and the strain was reduced to ensure an optimum torque (stress) response.

For data from both DMA and SAOS, Time Temperature Superposition (TTS) was used to form master curves. TTS is relevant to use when a material, to a good approximation, is controlled by a single characteristic time-scale. Such thermorheologically simple [111] behavior has been shown to be a good approximation for some LCE systems [1,78,84]. We initially investigate the validity of TTS by plotting $\tan(\delta)$ vs. $|G^*|$ (or $|E^*|$) in a so-called van Gurp Palmen (vGP) plot (see Supplementary Materials) [79]. The vGP representation removes all explicit time-dependence from the unshifted rheological data and therefore, shows if accurate TTS using frequency shifts is possible. TTS was carried out by frequency (horizontal) shifting the G' and G'', or E' and E'', data to form a single master curve, yielding a frequency shift factor at each measured temperature ($a_T(T)$). The horizontal shift factors $a_T(T)$ are thus given by:

$$\alpha_T = \frac{\tau(T)}{\tau(T_0)} = \frac{\omega(T_0)}{\omega(T)}, \tag{11}$$

where $\tau(T)$ is the timescale of the response measured at T, and $\tau(T_0)$ is the timescale of the response at a reference temperature, T_0. A reference temperature of T_0 = 40 °C was used for both the SAOS and DMA data. When performing TTS analysis, a vertical shift factor, $b_T(T)$ is sometimes required, e.g., to account for density changes in the material. However, for our LCE data, it was not necessary to invoke a vertical shift factor to describe our data, which is supported by the vGP analysis. To determine $\tau_\alpha(T)$ from the SAOS data, the characteristic time-scale of the α relaxation at T_0 was determined from the peak in G''. The shift factors at all other temperatures were used to determine $\tau_\alpha(T)$.

The effect of applied strain on E' and E'' on the nematic LCE was investigated using DMA. The sample was held under strain between the RSAII tension clamps and stress-relaxed for 2 min. A 1 Hz oscillatory strain with an amplitude of 0.1% was applied at 23 °C to the strained nematic LCE and E^* was determined. The sample is strained further, allowed to stress relax, and dynamic measurements are performed again. This allowed one to investigate E' and E'' as a function of applied strain up until sample failure. To correct for changes in cross-sectional area due to the applied strain, an approximation of constant volume was applied; this assumption is known to be a good approximation for the nematic LCE [18].

3.4. *Differential Scanning Calorimetry and Modulated Differential Scanning Calorimetry*

Conventional Differential Scanning Calorimetry (DSC), as well as modulated Differential Scanning Calorimetry (m-DSC), was employed to investigate the glass transitions of the isotropic and nematic LCEs. A TA Instruments Q2000 was used to perform both DSC and m-DSC. Approximately 10 mg of LCE sample was cut into 5 mm circular pieces and stacked into hermetic DSC pans. Conventional DSC was performed using a heating/cooling rate of 10 K/min. Each sample was held at 100 °C for 5 min to remove any thermal history. DSC heating/cooling runs were performed for 3 cycles between −70 °C and 80 °C, and the glass transition temperatures (T_g) were defined from the inflection points of the heat flow on the 2nd cooling cycle. m-DSC was performed using a linear heating ramp with a superimposed sinusoidal heating/cooling profile. A temperature amplitude of 1.2 K and a modulation period of 60 s was used with an underlying heating rate of 0.83 K/min across the glass transition regions. T_g values were defined from the inflection point of the reversing heat flow A modulation period of 60 s corresponds to an α relaxation time scale of $\tau_\alpha = 60/2\pi$ = 9.56 s.

3.5. Determining the Correlation Volume from Broadband Dielectric Relaxation Data

Using our BDS data for the isotropic LCE, we define a normalized dynamic susceptibility $\chi(f,T) = [\varepsilon'(f,T) - \varepsilon_\infty(T)]/\Delta\varepsilon(T)$, where $\varepsilon'(f,T)$, $\varepsilon_\infty(T)$ and $\Delta\varepsilon(T)$ are determined from the HN α relaxation contribution to the experimental data. Using the approximation discussed in the main text of references [89,94], it has been demonstrated that the number of correlated molecular units (molecules for a liquid or, e.g., monomers for a polymer), $N_{\text{corr},4}$ can be determined as:

$$N_{\text{corr},4}(T) \approx \frac{k_B}{\Delta C_p^{mol}} T^2 \left\{ \frac{max}{f} |\partial\chi(f,T)/\partial T| \right\}^2, \tag{12}$$

where k_B is the Bolzmann constant and ΔC_p^{mol} is the isobaric configurational heat capacity per chosen molecular unit, directly related to the α relaxation. More precisely, Equation (15) is a lower bound for $N_{\text{corr},4}$ but has been demonstrated to constitute a good approximation for both polymeric and non-polymeric systems [89,95]. ΔC_p^{mol} can be converted to a specific heat: $\Delta c_p = \Delta C_p^{mol} \cdot N_A/m_0$, where N_A is the Avogadro constant and m_0 is the molar weight of the chosen molecular unit. Moreover, we can define a correlation volume as:

$$V_{\text{corr, }4} = N_{\text{corr},4} \cdot \frac{m_0}{\rho \cdot N_A}, \tag{13}$$

where ρ is the volumetric mass density. Thus, in conclusion, we find that:

$$V_{\text{corr},4}(T) \approx \frac{k_B}{\Delta c_p \rho} T^2 \left\{ \frac{max}{f} |\partial\chi(f,T)/\partial T| \right\}^2, \tag{14}$$

where Δc_p is the isobaric configurational specific heat associated with the α relaxation. Here, we determine Δc_p directly from the measured specific heat step at T_g, We thus ignore the weak T-dependence of Δc_p. We also note that to estimate the contribution to the heat capacity arising from configurational degrees of freedom, it is common to subtract the contribution from the corresponding crystal, where mainly vibrational degrees of freedom contribute. However, since this is not possible for our material, we use the glassy state as a reference where relatively few configurational rearrangements take place and the vibrational contribution is dominating. The mass density, ρ, of the LCE is determined by measuring the mass of an LCE sample using a Mettler Toledo ME weighing scale, and by measuring the corresponding dimensions of the LCE sample using a Mituoyo Quantamike IP65 digital micrometer and calculating the volume of the sample. The density of the isotropic sample is 1300 ± 200 kg/m^3 at 20 °C and we ignore any weak temperature dependence in ρ. We note that the primary contribution to $V_{\text{corr},4}(T)$ arise from variations in $\tau_\alpha(T)$, which justifies our approximation of T-independent Δc_p and ρ. We determined $\partial\chi(f,T)/\partial T$ using finite differences as:

$$\frac{\partial\chi(f,T)}{\partial T} \approx \frac{\left[\chi\left(f, T + \frac{\Delta T}{2}\right) - \chi\left(f, T - \frac{\Delta T}{2}\right)\right]}{\Delta T}. \tag{15}$$

The dielectric data were recorded for temperatures T recorded in steps of 2 °C. For calculating the finite differences at each T, we use $\Delta T = 0.05$ °C in Equation (15); for this very small ΔT, we assume that TTS is valid and use the HN parameters determined for T to determine $\chi(f, T \pm \Delta T/2)$, except for τ_α which was determined from the VFT fit to the BDS data.

4. Conclusions

In this article, the molecular relaxations of the isotropic and nematic phases of an acrylate-based LCE, and their response to applied strain, have been determined using a combination of Broadband Dielectric Spectroscopy (BDS), Differential Scanning Calorimetry (DSC), Small Amplitude Oscillatory Shear Rheology (SAOS) and tensile Dynamic Mechanical Analysis (DMA). Due to its lack of phase transitions across a wide T-range, once polymerized, our chosen LCE constitutes an excellent model system for investigating the effects of nematic order on the glass-transition-related molecular relaxation behavior. Moreover, when strained, our LCE deforms via a mechanical Fréedericksz transition, demonstrating a molecular auxetic response, whose detailed origin is not well understood [18,19]. To better understand this behavior, we have investigated the effects of deformation on both the relaxation dynamics and the mechanical response of our nematic LCE.

For the quiescent LCEs, we demonstrate that both the isotropic and nematic LCE samples show a similar T-dependence of their characteristic α relaxation time-scales $\tau_\alpha(T)$ near the glass transition temperatures T_g; this is reflected in the similar T_g-values (T_g = 279 K and 278 K, as probed by DSC) and dynamic fragilities ($D = 5.4$ and $D = 5.1$; $m = 110$ and $m = 130$), for the two phases, which are both representing fragile glass-formers with fragility values consistent with those of polymers [23,24]. Importantly, for both phases, $\tau_\alpha(T)$ qualitatively changes T-dependence at a crossover $T^* \approx 333$ K. For $T > T^*$, $\tau_\alpha(T)$ of the nematic LCE is well described as Arrhenius, whereas the isotropic LCE is non-Arrhenius, but more Arrhenius-like (less fragile) above, than below, T^*. A similar change from non-Arrhenius to Arrhenius at a crossover temperature is often observed for liquid crystal side-chain polymers with nematic phases [42,46].

For comparison, a change in the T-dependence of $\tau_\alpha(T)$ is typically observed in glass-formers at a temperature $T_B \sim 1.2 - 1.6 \times T_g$ [30], which is comparable to the $T^* \sim T_g$, observed for our LCEs. However, for non-polymeric non-LC glass-formers, $\tau_\alpha(T)$ is typically less Arrhenius-like (more fragile) above, than below, T_B and the change in T-dependence often disappears all together for polymeric systems [31]. Thus, the behavior observed for many nematic systems displays both similarities and clear differences to that of non-LC systems. For our LCE, we stress that the change of phase does not significantly affect the value of T^*, but it strongly affects how Arrhenius-like the T-dependence of $\tau_\alpha(T)$ is for $T > T^*$. We propose that these observations could be related to the existence of pre-transitional phenomena in the form of nematic regions in our 'isotropic' LCE, which are locked in by the polymerization. We estimate the size of these regions to be $\sim$1 nm, and demonstrate, by analysis of our BDS data on the isotropic LCE, that correlated segmental motions on similar length-scales are involved in the α relaxation near the crossover at T^*.

The extent of decoupling between the ionic conductivity and the structural α relaxation was investigated and we found that the ionic conductivity was significantly more decoupled from the α relaxation in the nematic than in the isotropic LCE. We suggest that differences in dynamic heterogeneity in the two LCE phases could be an important contributing factor in these results. Furthermore, the structural anisotropy induced by the nematic order could in itself play a role in driving the observed decoupling. In addition to the α relaxation, two further dielectrically active relaxations (β and γ) are observed in both LCE phases. They both demonstrate Arrhenius behavior within the glassy state and are not significantly affected by the LCE phase; we assign these to motions of the mesogen side-chain around its long axis (β), and motions of the alkyl spacer (γ), respectively [22,40]. In addition, a rheology investigation of the isotropic LCE provided evidence for both a Rouse-like mode contribution on time-scales slower than the α relaxation, and an additional relaxation contribution suggested to be due either to free chains moving within the elastomer network, or to the motion of pendant network chains [85–87].

Finally, to investigate the origin of molecular auxetic behavior in the nematic LCE, the effects of the applied strain are determined. The complex Young's modulus, E^*, and the α relaxation time τ_α are measured for varying applied strains using BDS and DMA. We find that E', E'', and $\tau_\alpha(T)$ remain relatively unchanged for increasing strain, ϵ_t, until $\epsilon_t > 0.55$, where both metrics increase; we associate this increase with a strain-induced increase of constraints on the polymer backbone and the finite extensibility of the network [9,105,109]. Importantly, the mechanical Fréedericksz transition, and the molecular auxetic response, both occur near the region of applied strain, where E', E'', and τ_α increase. Based on the previous suggestion that the molecular auxetic response is linked to the emergence of biaxial order, in turn caused by out-of-plane rotations of mesogenic units (above an applied threshold strain) [19], we suggest that the rotation of the mesogen units is itself driven by the growth of constraints imposed by the increasing strain.

Supplementary Materials: The following are available online, Figure S1: Full dielectric spectra for the isotropic and nematic LCE as a function of temperature, Figure S2: Dielectric spectra in the dielectric modulus and derivative representation, Figure S3: Van Gurp–Palmen graphs for the shear rheology and DMA, Figure S4: Unshifted shear rheology data as a function of temperature, Figure S5: Unshifted DMA data as a function of temperature, Figure

S6: DSC and m-DSC traces on the isotropic and nematic LCE samples, Figure S7: normalised BDS ε'' data taken at 23 °C as a function of applied true strain for the low and high strain sample.

Author Contributions: Conceptualization, T.R., J.M. and H.F.G.; experimental investigation, T.R. and M.R.; writing—review and editing, T.R., M.R., J.M. and H.F.G.; figure preparation, T.R. and M.R. All authors have read and agreed to the published version of the manuscript.

Funding: M.R., H.F.G. and J.M. thank the EPSRC (EP/M009521/1) for support. T.R. thanks the University of Leeds Alumni Fund.

Institutional Review Board Statement: Not applicable.

Informed Consent Statement: Not applicable.

Data Availability Statement: Data made available at https://doi.org/10.5518/1074 (latest accessed on 29 November 2021).

Acknowledgments: The authors thanks D. L. Baker and Z. Zhang of the university for discussions relating to BDS. The authors thank D. L. Baker for discussion on glassy behavior. The authors thank J. Hobbs for discussion on nematic glasses.

Conflicts of Interest: The authors declare no conflict of interest.

Sample Availability: Not available.

References

1. Warner, M.; Terentjev, E.M. *Liquid Crystal Elastomers*; Oxford University Press: Oxford, UK, 2007.
2. Zeng, H.; Wani, O.M.; Wasylczyk, P.; Priimagi, A. Light-Driven, Caterpillar-Inspired Miniature Inching Robot. *Macromol. Rapid Commun.* **2018**, *39*, 1700224. [CrossRef]
3. Boothby, J.M.; Kim, H.; Ware, T.H. Shape Changes in Chemoresponsive Liquid Crystal Elastomers. *Sens. Actuators B Chem.* **2017**, *240*, 511–518. [CrossRef]
4. Mistry, D.; Nikkhou, M.; Raistrick, T.; Hussain, M.; Jull, E.I.L.; Gleeson, H.F. Isotropic Liquid Crystal Elastomers as Exceptional Photoelastic Strain Sensors. *Macromolecules* **2020**, *53*, 3709–3718. [CrossRef]
5. Kundler, I.; Finkelmann, H. Strain-Induced Director Reorientation in Nematic Liquid Single Crystal Elastomers. *Macromol. Rapid Commun.* **1995**, *16*, 679–686. [CrossRef]
6. Hussain, M.; Jull, E.I.L.; Mandle, R.J.; Raistrick, T.; Hine, P.J.; Gleeson, H.F. Liquid Crystal Elastomers for Biological Applications. *Nanomaterials* **2021**, *11*, 813. [CrossRef] [PubMed]
7. Merkel, D.R.; Traugutt, N.A.; Visvanathan, R.; Yakacki, C.M.; Frick, C.P. Thermomechanical Properties of Monodomain Nematic Main-Chain Liquid Crystal Elastomers. *Soft Matter* **2018**, *14*, 6024–6036. [CrossRef]
8. Yin, H.; Chapala, P.; Bermeshev, M.; Schönhals, A.; Böhning, M. Molecular Mobility and Physical Aging of a Highly Permeable Glassy Polynorbornene as Revealed by Dielectric Spectroscopy. *ACS Macro Lett.* **2017**, *6*, 813–818. [CrossRef]
9. Lee, H.; Fragiadakis, D.; Martin, D.; Milne, A.; Milne, J.; Runt, J. Dynamics of Uniaxially Oriented Elastomers Using Broadband Dielectric Spectroscopy. *Macromolecules* **2010**, *43*, 3125–3127. [CrossRef]
10. Riggleman, R.A.; Lee, H.-N.; Ediger, M.D.; Pablo, J.J. de Heterogeneous Dynamics during Deformation of a Polymer Glass. *Soft Matter* **2010**, *6*, 287–291. [CrossRef]
11. Lee, H.-N.; Paeng, K.; Swallen, S.F.; Ediger, M.D.; Stamm, R.A.; Medvedev, G.A.; Caruthers, J.M. Molecular Mobility of Poly(Methyl Methacrylate) Glass during Uniaxial Tensile Creep Deformation. *J. Polym. Sci. Part B Polym. Phys.* **2009**, *47*. [CrossRef]
12. Bending, B.; Christison, K.; Ricci, J.; Ediger, M.D. Measurement of Segmental Mobility during Constant Strain Rate Deformation of a Poly(Methyl Methacrylate) Glass. *Macromolecules* **2014**, *47*, 800–806. [CrossRef]
13. Ratner, M.A.; Johansson, P.; Shriver, D.F. Polymer Electrolytes: Ionic Transport Mechanisms and Relaxation Coupling. *MRS Bull.* **2000**, *25*, 31–37. [CrossRef]
14. Mistry, D.; Gleeson, H.F. Mechanical Deformations of a Liquid Crystal Elastomer at Director Angles between 0° and 90°: Deducing an Empirical Model Encompassing Anisotropic Nonlinearity. *J. Polym. Sci. Part B Polym. Phys.* **2019**, *57*, 1367–1377. [CrossRef]
15. Drozd-Rzoska, A. Heterogeneity-Related Dynamics in Isotropic n-Pentylcyanobiphenyl. *Phys. Rev. E* **2006**, *73*, 022501. [CrossRef] [PubMed]
16. Drozd-Rzoska, A. Glassy Dynamics of Liquid Crystalline 4′-n-Pentyl-4-Cyanobiphenyl in the Isotropic and Supercooled Nematic Phases. *J. Chem. Phys.* **2009**, *130*, 234910. [CrossRef] [PubMed]
17. Drozd-Rzoska, A.; Rzoska, S.J.; Paluch, M.; Pawlus, S.; Zioło, J.; Santangelo, P.G.; Roland, C.M.; Czupryński, K.; Dąbrowski, R. Mode Coupling Behavior in Glass-Forming Liquid Crystalline Isopentylcyanobiphenyl. *Phys. Rev. E* **2005**, *71*, 011508. [CrossRef]
18. Mistry, D.; Connell, S.D.; Mickthwaite, S.L.; Morgan, P.B.; Clamp, J.H.; Gleeson, H.F. Coincident Molecular Auxeticity and Negative Order Parameter in a Liquid Crystal Elastomer. *Nat. Commun.* **2018**, *9*, 5095. [CrossRef] [PubMed]

19. Raistrick, T.; Zhang, Z.; Mistry, D.; Mattsson, J.; Gleeson, H.F. Understanding the Physics of the Auxetic Response in a Liquid Crystal Elastomer. *Phys. Rev. Res.* **2021**, *3*, 023191. [CrossRef]
20. Mistry, D.; Morgan, P.B.; Clamp, J.H.; Gleeson, H.F. New Insights into the Nature of Semi-Soft Elasticity and "Mechanical-Fréedericksz Transitions" in Liquid Crystal Elastomers. *Soft Matter* **2018**, *14*, 1301–1310. [CrossRef]
21. Debenedetti, P.G.; Stillinger, F.H. Supercooled Liquids and the Glass Transition. *Nature* **2001**, *410*, 259–267. [CrossRef]
22. Kremer, F.; Schönhals, A. (Eds.) *Broadband Dielectric Spectroscopy*; Springer: Berlin/Heidelberg, Germany, 2003; ISBN 978-3-642-62809-2.
23. Roland, C.M. Relaxation Phenomena in Vitrifying Polymers and Molecular Liquids. *Macromolecules* **2010**, *43*, 7875–7890. [CrossRef]
24. Casalini, R.; Roland, C.M.; Capaccioli, S. Effect of Chain Length on Fragility and Thermodynamic Scaling of the Local Segmental Dynamics in Poly(Methylmethacrylate). *J. Chem. Phys.* **2007**, *126*, 184903. [CrossRef] [PubMed]
25. Ngai, K.L.; Lunkenheimer, P. Nature and Properties of the Johari–Goldstein β-Relaxation in the Equilibrium Liquid State of a Class of Glass-Formers. *J. Chem. Phys.* **2001**, *115*, 9. [CrossRef]
26. Ngai, K.; Capaccioli, S. Relation between the Activation Energy of the Johari-Goldstein β Relaxation and Tg of Glass Formers. *Phys. Rev. E* **2004**, *69*, 031501. [CrossRef] [PubMed]
27. Kudlik, A.; Tschirwitz, C.; Benkhof, S.; Blochowicz, T.; Rössler, E. Slow Secondary Relaxation Process in Supercooled Liquids. *EPL* **1997**, *40*, 649. [CrossRef]
28. Paluch, M.; Roland, C.M.; Pawlus, S.; Zioło, J.; Ngai, K.L. Does the Arrhenius Temperature Dependence of the Johari-Goldstein Relaxation Persist above Tg? *Phys. Rev. Lett.* **2003**, *91*, 115701. [CrossRef] [PubMed]
29. Wang, L.-M.; Angell, C.A.; Richert, R. Fragility and Thermodynamics in Nonpolymeric Glass-Forming Liquids. *J. Chem. Phys.* **2006**, *125*, 074505. [CrossRef] [PubMed]
30. Novikov, V.N.; Sokolov, A.P. Universality of the Dynamic Crossover in Glass-Forming Liquids: A "Magic" Relaxation Time. *Phys. Rev. E* **2003**, *67*, 031507. [CrossRef] [PubMed]
31. Stickel, F.; Fischer, E.W.; Richert, R. Dynamics of Glass-forming Liquids. II. Detailed Comparison of Dielectric Relaxation, Dc-conductivity, and Viscosity Data. *J. Chem. Phys.* **1996**, *104*, 2043–2055. [CrossRef]
32. Goldstein, M. Viscous Liquids and the Glass Transition: A Potential Energy Barrier Picture. *J. Chem. Phys.* **1969**, *51*, 3728–3739. [CrossRef]
33. Baker, D.L.; Reynolds, M.; Masurel, R.; Olmsted, P.D.; Mattsson, J. Chain-Length, Flexibility and the Glass Transition of Polymers. *arXiv* **2019**, arXiv:1911.13278.
34. Colmenero, J. Are Polymers Standard Glass-Forming Systems? The Role of Intramolecular Barriers on the Glass-Transition Phenomena of Glass-Forming Polymers. *J. Phys. Condens. Matter* **2015**, *27*, 103101. [CrossRef] [PubMed]
35. Mattsson, J.; Bergman, R.; Jacobsson, P.; Börjesson, L. Chain-Length-Dependent Relaxation Scenarios in an Oligomeric Glass-Forming System: From Merged to Well-Separated α and β Loss Peaks. *Phys. Rev. Lett.* **2003**, *90*, 075702. [CrossRef]
36. Schroeder, M.J.; Roland, C.M. Segmental Relaxation in End-Linked Poly(Dimethylsiloxane) Networks. *Macromolecules* **2002**, *35*, 2676–2681. [CrossRef]
37. Lal, S.; Tripathi, S.K.; Sood, N.; Khosla, S. Study of Dielectric Parameters of Liquid Crystal Elastomer. *Liq. Cryst.* **2014**, *41*, 1402–1409. [CrossRef]
38. Brömmel, F.; Stille, W.; Finkelmann, H.; Hoffmann, A. Molecular Dynamics and Biaxiality of Nematic Polymers and Elastomers. *Soft Matter* **2011**, *7*, 2387–2401. [CrossRef]
39. Shenoy, D.; Filippov, S.; Aliev, F.; Keller, P.; Thomsen, D.; Ratna, B. Vogel-Fulcher Dependence of Relaxation Rates in a Nematic Monomer and Elastomer. *Phys. Rev. E* **2000**, *62*, 8100–8105. [CrossRef] [PubMed]
40. Schönhals, A. Hans-eckartcarius Dielectric Properties of Thermotropic Polymer Liquid Crystals. *Int. J. Polym. Mater.* **2000**, *45*, 239–276. [CrossRef]
41. Zentel, R.; Strobl, G.R.; Ringsdorf, H. Dielectric Relaxation of Liquid Crystalline Polyacrylates and Polymethacrylates. *Macromolecules* **1985**, *18*, 960–965. [CrossRef]
42. Schönhals, A.; Wolff, D.; Springer, J. Temperature Dependence of the Relaxation Rates of α and δ Relaxation in Liquid-Crystalline Side-Group Polymethacrylates. *Macromolecules* **1998**, *31*, 9019–9025. [CrossRef]
43. Salehli, F.; Yildiz, S.; Ozbek, H.; Uykur, E.; Gursel, Y.H.; Durmaz, Y.Y. Peculiarities of δ- and α-Relaxations in Thermotropic Side Chain Liquid Crystalline Polymers with and without Nematic Reentrant Phase. *Polymer* **2010**, *51*, 1450–1456. [CrossRef]
44. Ngai, K.L.; Schönhals, A. Interpretation of the Observed Influence of Mesophase Structures on the β-Relaxation in Side Chain Liquid Crystal Polymers. *J. Polym. Sci. Part B Polym. Phys.* **1998**, *36*, 1927–1934. [CrossRef]
45. Vallerien, S.U.; Kremer, F.; Boeffel, C. Broadband Dielectric Spectroscopy on Side Group Liquid Crystal Polymers. *Liq. Cryst.* **1989**, *4*, 79–86. [CrossRef]
46. Turky, G.; Wolff, D.; Schönhals, A. Confinement Effects on the Molecular Dynamics of Liquid-Crystalline Polymethacrylates—A Broadband Dielectric Spectroscopy Study. *Macromol. Chem. Phys.* **2012**, *213*, 2420–2431. [CrossRef]
47. Verwey, G.C.; Warner, M.; Terentjev, E.M. Elastic Instability and Stripe Domains in Liquid Crystalline Elastomers. *J. Phys. II Fr.* **1996**, *6*, 1273–1290. [CrossRef]
48. Stein, P.; Aßfalg, N.; Finkelmann, H.; Martinoty, P. Shear Modulus of Polydomain, Mono-Domain and Non-Mesomorphic Side-Chain Elastomers: Influence of the Nematic Order. *Eur. Phys. J. E* **2001**, *4*, 255–262. [CrossRef]

49. Rogez, D.; Francius, G.; Finkelmann, H.; Martinoty, P. Shear Mechanical Anisotropy of Side Chain Liquid-Crystal Elastomers: Influence of Sample Preparation. *Eur. Phys. J. E* **2006**, *20*, 369–378. [CrossRef]
50. Weilepp, J.; Zanna, J.J.; Assfalg, N.; Stein, P.; Hilliou, L.; Mauzac, M.; Finkelmann, H.; Brand, H.R.; Martinoty, P. Rheology of Liquid Crystalline Elastomers in Their Isotropic and Smectic A State. *Macromolecules* **1999**, *32*, 4566–4574. [CrossRef]
51. Roberts, P.M.S.; Mitchell, G.R.; Davis, F.J. A Single Director Switching Mode for Monodomain Liquid Crystal Elastomers. *J. Phys. II Fr.* **1997**, *7*, 1337–1351. [CrossRef]
52. Bladon, P.; Warner, M.; Terentjev, E.M. Orientational Order in Strained Nematic Networks. *Macromolecules* **1994**, *27*, 7067–7075. [CrossRef]
53. Wübbenhorst, M.; van Turnhout, J. Analysis of Complex Dielectric Spectra. I. One-Dimensional Derivative Techniques and Three-Dimensional Modelling. *J. Non-Cryst. Solids* **2002**, *305*, 40–49. [CrossRef]
54. Gedde, U.W.; Liu, F.; Hult, A.; Sahlén, F.; Boyd, R.H. Dielectric Relaxation of Liquid Crystalline Side-Chain Poly(Vinyl Ether)s. *Polymer* **1994**, *35*, 2056–2062. [CrossRef]
55. Martínez-Felipe, A.; Santonja-Blasco, L.; Badia, J.D.; Imrie, C.T.; Ribes-Greus, A. Characterization of Functionalized Side-Chain Liquid Crystal Methacrylates Containing Nonmesogenic Units by Dielectric Spectroscopy. *Ind. Eng. Chem. Res.* **2013**, *52*, 8722–8731. [CrossRef]
56. Stickel, F.; Fischer, E.W.; Richert, R. Dynamics of Glass-forming Liquids. I. Temperature-derivative Analysis of Dielectric Relaxation Data. *J. Chem. Phys.* **1995**, *102*, 6251–6257. [CrossRef]
57. Schick, C.; Sukhorukov, D.; Schönhals, A. Comparison of the Molecular Dynamics of a Liquid Crystalline Side Group Polymer Revealed from Temperature Modulated DSC and Dielectric Experiments in the Glass Transition Region. *Macromol. Chem. Phys.* **2001**, *202*, 1398–1404. [CrossRef]
58. Zhang, L.; Yao, W.; Gao, Y.; Zhang, C.; Yang, H. Polysiloxane-Based Side Chain Liquid Crystal Polymers: From Synthesis to Structure–Phase Transition Behavior Relationships. *Polymers* **2018**, *10*, 794. [CrossRef] [PubMed]
59. Wang, Y.; Fan, F.; Agapov, A.L.; Yu, X.; Hong, K.; Mays, J.; Sokolov, A.P. Design of Superionic Polymers—New Insights from Walden Plot Analysis. *Solid State Ion.* **2014**, *262*, 782–784. [CrossRef]
60. Agapov, A.L.; Sokolov, A.P. Decoupling Ionic Conductivity from Structural Relaxation: A Way to Solid Polymer Electrolytes? *Macromolecules* **2011**, *44*, 4410–4414. [CrossRef]
61. Bocharova, V.; Sokolov, A.P. Perspectives for Polymer Electrolytes: A View from Fundamentals of Ionic Conductivity. *Macromolecules* **2020**, *53*, 4141–4157. [CrossRef]
62. Funahashi, M.; Shimura, H.; Yoshio, M.; Kato, T. Functional Liquid-Crystalline Polymers for Ionic and Electronic Conduction. In *Liquid Crystalline Functional Assemblies and Their Supramolecular Structures*; Kato, T., Ed.; Structure and Bonding; Springer: Berlin/Heidelberg, Germany, 2008; pp. 151–179. ISBN 978-3-540-77867-7.
63. Kishimoto, K.; Suzawa, T.; Yokota, T.; Mukai, T.; Ohno, H.; Kato, T. Nano-Segregated Polymeric Film Exhibiting High Ionic Conductivities. *J. Am. Chem. Soc.* **2005**, *127*, 15618–15623. [CrossRef] [PubMed]
64. Feng, C.; Rajapaksha, C.P.H.; Cedillo, J.M.; Piedrahita, C.; Cao, J.; Kaphle, V.; Lüssem, B.; Kyu, T.; Jákli, A. Electroresponsive Ionic Liquid Crystal Elastomers. *Macromol. Rapid Commun.* **2019**, *40*, 1900299. [CrossRef] [PubMed]
65. Chambers, M.; Zalar, B.; Remškar, M.; Kovač, J.; Finkelmann, H.; Žumer, S. Investigations on an Integrated Conducting Nanoparticle–Liquid Crystal Elastomer Layer. *Nanotechnology* **2007**, *18*, 415706. [CrossRef]
66. Wang, Y.; Agapov, A.L.; Fan, F.; Hong, K.; Yu, X.; Mays, J.; Sokolov, A.P. Decoupling of Ionic Transport from Segmental Relaxation in Polymer Electrolytes. *Phys. Rev. Lett.* **2012**, *108*, 088303. [CrossRef]
67. Do, C.; Lunkenheimer, P.; Diddens, D.; Götz, M.; Weiß, M.; Loidl, A.; Sun, X.-G.; Allgaier, J.; Ohl, M. Li^+ Transport in Poly(Ethylene Oxide) Based Electrolytes: Neutron Scattering, Dielectric Spectroscopy, and Molecular Dynamics Simulations. *Phys. Rev. Lett.* **2013**, *111*, 018301. [CrossRef]
68. Golodnitsky, D.; Strauss, E.; Peled, E.; Greenbaum, S. Review—On Order and Disorder in Polymer Electrolytes. *J. Electrochem. Soc.* **2015**, *162*, A2551. [CrossRef]
69. Dyre, J.C. Colloquium: The Glass Transition and Elastic Models of Glass-Forming Liquids. *Rev. Mod. Phys.* **2006**, *78*, 953–972. [CrossRef]
70. Price, H.C.; Mattsson, J.; Murray, B.J. Sucrose Diffusion in Aqueous Solution. *Phys. Chem. Chem. Phys.* **2016**, *18*, 19207–19216. [CrossRef] [PubMed]
71. Pollack, G.L. Atomic Test of the Stokes-Einstein Law: Diffusion and Solubility of Xe. *Phys. Rev. A* **1981**, *23*, 2660–2663. [CrossRef]
72. Ediger, M.D.; Harrowell, P.; Yu, L. Crystal Growth Kinetics Exhibit a Fragility-Dependent Decoupling from Viscosity. *J. Chem. Phys.* **2008**, *128*, 034709. [CrossRef] [PubMed]
73. Ediger, M.D. Can Density or Entropy Fluctuations Explain Enhanced Translational Diffusion in Glass-Forming Liquids? *J. Non-Cryst. Solids* **1998**, *235–237*, 10–18. [CrossRef]
74. Ediger, M.D.; Angell, C.A.; Nagel, S.R. Supercooled Liquids and Glasses. *J. Phys. Chem.* **1996**, *100*, 13200–13212. [CrossRef]
75. Yu, H.B.; Wang, W.H.; Bai, H.Y.; Samwer, K. The β-Relaxation in Metallic Glasses. *Natl. Sci. Rev.* **2014**, *1*, 429–461. [CrossRef]
76. Drozd-Rzoska, A.; Starzonek, S.; Rzoska, S.J.; Kralj, S. Nanoparticle-Controlled Glassy Dynamics in Nematogen-Based Nanocolloids. *Phys. Rev. E* **2019**, *99*, 052703. [CrossRef] [PubMed]
77. Andreev, Y.G.; Bruce, P.G. Polymer Electrolyte Structure and Its Implications. *Electrochim. Acta* **2000**, *45*, 1417–1423. [CrossRef]

78. Zanna, J.J.; Stein, P.; Marty, J.D.; Mauzac, M.; Martinoty, P. Influence of Molecular Parameters on the Elastic and Viscoelastic Properties of Side-Chain Liquid Crystalline Elastomers. *Macromolecules* **2002**, *35*, 5459–5465. [CrossRef]
79. van Gurp, M.; Palmen, J. Time-Temperature Superposition for Polymeric Blends. *Rheol. Bull.* **1998**, *67*, 5–8.
80. Ren, W.; McMullan, P.J.; Griffin, A.C. Stress–Strain Behavior in Main Chain Liquid Crystalline Elastomers: Effect of Crosslinking Density and Transverse Rod Incorporation on "Poisson's Ratio". *Phys. Status Solidi* **2009**, *246*, 2124–2130. [CrossRef]
81. Ren, W.; McMullan, P.J.; Griffin, A.C. Poisson's Ratio of Monodomain Liquid Crystalline Elastomers. *Macromol. Chem. Phys.* **2008**, *209*, 1896–1899. [CrossRef]
82. Vandoolaeghe, W.L.; Terentjev, E.M. Constrained Rouse Model of Rubber Viscoelasticity. *J. Chem. Phys.* **2005**, *123*, 034902. [CrossRef] [PubMed]
83. Rubinstein, M.; Colby, R. (Eds.) *Polymer Physics*; Oxford University Press: Oxford, UK, 2003.
84. Gallani, J.L.; Hilliou, L.; Martinoty, P.; Doublet, F.; Mauzac, M. Mechanical Behavior of Side-Chain Liquid Crystalline Networks. *J. Phys. II* **1996**, *6*, 443–452. [CrossRef]
85. Urayama, K.; Yokoyama, K.; Kohjiya, S. Viscoelastic Relaxation of Guest Linear Poly(Dimethylsiloxane) in End-Linked Poly(Dimethylsiloxane) Networks. *Macromolecules* **2001**, *34*, 4513–4518. [CrossRef]
86. Yamazaki, H.; Takeda, M.; Kohno, Y.; Ando, H.; Urayama, K.; Takigawa, T. Dynamic Viscoelasticity of Poly(Butyl Acrylate) Elastomers Containing Dangling Chains with Controlled Lengths. *Macromolecules* **2011**, *44*, 8829–8834. [CrossRef]
87. Agudelo, D.C.; Roth, L.E.; Vega, D.A.; Vallés, E.M.; Villar, M.A. Dynamic Response of Transiently Trapped Entanglements in Polymer Networks. *Polymer* **2014**, *55*, 1061–1069. [CrossRef]
88. Jakobsen, B.; Niss, K.; Maggi, C.; Olsen, N.B.; Christensen, T.; Dyre, J.C. Beta Relaxation in the Shear Mechanics of Viscous Liquids: Phenomenology and Network Modeling of the Alpha-Beta Merging Region. *J. Non-Cryst. Solids* **2011**, *357*, 267–273. [CrossRef]
89. Dalle-Ferrier, C.; Thibierge, C.; Alba-Simionesco, C.; Berthier, L.; Biroli, G.; Bouchaud, J.-P.; Ladieu, F.; L'Hôte, D.; Tarjus, G. Spatial Correlations in the Dynamics of Glassforming Liquids: Experimental Determination of Their Temperature Dependence. *Phys. Rev. E* **2007**, *76*, 041510. [CrossRef]
90. Tracht, U.; Wilhelm, M.; Heuer, A.; Feng, H.; Schmidt-Rohr, K.; Spiess, H.W. Length Scale of Dynamic Heterogeneities at the Glass Transition Determined by Multidimensional Nuclear Magnetic Resonance. *Phys. Rev. Lett.* **1998**, *81*, 2727–2730. [CrossRef]
91. Donth, E. The Size of Cooperatively Rearranging Regions at the Glass Transition. *J. Non-Cryst. Solids* **1982**, *53*, 325–330. [CrossRef]
92. Stevenson, J.D.; Schmalian, J.; Wolynes, P.G. The Shapes of Cooperatively Rearranging Regions in Glass-Forming Liquids. *Nat. Phys.* **2006**, *2*, 268–274. [CrossRef]
93. Hempel, E.; Hempel, G.; Hensel, A.; Schick, C.; Donth, E. Characteristic Length of Dynamic Glass Transition near T_g for a Wide Assortment of Glass-Forming Substances. *J. Phys. Chem. B* **2000**, *104*, 2460–2466. [CrossRef]
94. Berthier, L. Direct Experimental Evidence of a Growing Length Scale Accompanying the Glass Transition. *Science* **2005**, *310*, 1797–1800. [CrossRef]
95. Capaccioli, S.; Ruocco, G.; Zamponi, F. Dynamically Correlated Regions and Configurational Entropy in Supercooled Liquids. *J. Phys. Chem. B* **2008**, *112*, 10652–10658. [CrossRef] [PubMed]
96. Krich, J.J.; Romanowsky, M.B.; Collings, P.J. Correlation Length and Chirality of the Fluctuations in the Isotropic Phase of Nematic and Cholesteric Liquid Crystals. *Phys. Rev. E* **2005**, *71*, 051712. [CrossRef]
97. Brand, H.R.; Kawasaki, K. On the Macroscopic Consequences of Frozen Order in Liquid Single Crystal Elastomers. *Macromol. Rapid Commun.* **1994**, *15*, 251–257. [CrossRef]
98. Petridis, L.; Terentjev, E.M. Nematic-Isotropic Transition with Quenched Disorder. *Phys. Rev. E* **2006**, *74*, 051707. [CrossRef]
99. Rzoska, S.J.; Drozd-Rzoska, A.; Mukherjee, P.K.; Lopez, D.O.; Martinez-Garcia, J.C. Distortion-Sensitive Insight into the Pretransitional Behavior of 4-n-Octyloxy-4′-Cyanobiphenyl (8OCB). *J. Phys. Condens. Matter* **2013**, *25*, 245105. [CrossRef] [PubMed]
100. Rzoska, S.J.; Starzonek, S.; Łoś, J.; Drozd-Rzoska, A.; Kralj, S. Dynamics and Pretransitional Effects in C60 Fullerene Nanoparticles and Liquid Crystalline Dodecylcyanobiphenyl (12CB) Hybrid System. *Nanomaterials* **2020**, *10*, 2343. [CrossRef] [PubMed]
101. Basu, S.; Hay, J.L.; Swindeman, J.E.; Oliver, W.C. Continuous Dynamic Analysis: Evolution of Elastic Properties with Strain. *MRS Commun.* **2014**, *4*, 25–29. [CrossRef]
102. Lv, Z.; Ren, J.; Lin, S.; Pei, Y.; Shao, Z.; Ling, S. Understanding the Continuous Dynamic Mechanical Behavior of Animal Silk. *Macromolecules* **2021**, *54*, 249–258. [CrossRef]
103. Meinecke, E.A. Dynamisches Verhalten von Elastomeren bei großen Dehnungen überlagerten kleinen Schwingungen. *Rheol. Acta* **1971**, *10*, 302–309. [CrossRef]
104. Suphadon, N.; Thomas, A.G.; Busfield, J.J.C. Viscoelastic Behavior of Rubber under a Complex Loading. *J. Appl. Polym. Sci.* **2009**, *113*, 693–699. [CrossRef]
105. Mason, P. The Viscoelastic Behavior of Rubber in Extension. *J. Appl. Polym. Sci.* **1959**, *1*, 63–69. [CrossRef]
106. Choi, T.; Fragiadakis, D.; Roland, C.M.; Runt, J. Microstructure and Segmental Dynamics of Polyurea under Uniaxial Deformation. *Macromolecules* **2012**, *45*, 3581–3589. [CrossRef]
107. Lee, H.-N.; Paeng, K.; Swallen, S.F.; Ediger, M.D. Direct Measurement of Molecular Mobility in Actively Deformed Polymer Glasses. *Science* **2009**, *323*, 231–234. [CrossRef]
108. Eyring, H. Viscosity, Plasticity, and Diffusion as Examples of Absolute Reaction Rates. *J. Chem. Phys.* **1936**, *4*, 283–291. [CrossRef]

109. Roland, C.M.; Hensel-Bielowka, S.; Paluch, M.; Casalini, R. Supercooled Dynamics of Glass-Forming Liquids and Polymers under Hydrostatic Pressure. *Rep. Prog. Phys.* **2005**, *68*, 1405–1478. [CrossRef]
110. Havriliak, S.; Havriliak, S.J. Results from an Unbiased Analysis of Nearly 1000 Sets of Relaxation Data. *J. Non-Cryst. Solids* **1994**, *172–174*, 297–310. [CrossRef]
111. Plazek, D.J. 1995 Bingham Medal Address: Oh, Thermorheological Simplicity, Wherefore Art Thou? *J. Rheol.* **1996**, *40*, 987–1014. [CrossRef]

molecules

MDPI

Article

Alcohol Selective Optical Sensor Based on Porous Cholesteric Liquid Crystal Polymer Networks

Tai-Yuan Yeh, Ming-Fu Liu, Ru-De Lin and Shug-June Hwang *

Department of Electro-Optical Engineering, National United University, Miao-Li 360, Taiwan;
taiyuan.ee10@nycu.edu.tw (T.-Y.Y.); zx74361296@gmail.com (M.-F.L.); xup6aul4eji3@gmail.com (R.-D.L.)
* Correspondence: june@nuu.edu.tw

Abstract: A responsive hydrogen-bonded cholesteric liquid crystal polymer (CLCP) film with controlled porosity was fabricated as an optical sensor to distinguish between methanol and ethanol in alcohol solutions. To facilitate responding the alcohols, porosity was generated by removing the nonreactive liquid crystal agent, and the hydrogen bridges of CLCP were broken. The sensitivities of CLCPs to ethanol and methanol were obtained by monitoring the wavelength shifts of the transmission spectrum at different alcohol concentrations and ratios of methanol/ethanol. Changes in the central wavelength of the CLCP network transmission spectrum allowed the methanol–ethanol ratio to be discriminated. A linear relationship between wavelength shift of CLCP networks and alcohol concentration was obtained experimentally, and the sensor characteristics were explored. The sensitivities of the CLCPs were 1.35 and 0.18 nm/% to ethanol and methanol, respectively. The sensing sensitivity of cholesteric networks to alcohol molecules increased as the methanol–ethanol ratio declined. Therefore, CLCP could act as a stimuli-responsive material to distinguish the concentrations of acetone and ethanol in mixed solutions. Furthermore, the impact of UV intensity for curing a CLC mixture on the sensing sensitivity to the different alcohol concentrations was also studied. The higher UV intensity could enhance the sensitivity to alcohol molecules and distinguishing ability between methanol and ethanol.

Keywords: cholesteric liquid crystal polymer; alcohol sensor; methanol detection; carboxylate polymer

1. Introduction

Methanol and ethanol are colorless, flammable, and easily soluble in water, and are commonly used in medical sterilization, chemical synthesis, and industrial fuels. Since these two alcohol solutions are easy to obtain and the price of methanol is lower than that of ethanol, some unscrupulous manufacturers maliciously make alcoholic beverages mixed with methanol. However, if methanol is ingested, inhaled, or absorbed by the skin, it can cause irreversible tissue injury, especially to the eyes and nervous system, and even death. To effectively prevent the threat of methanol to human health, the illegal use of methanol as fake ethanol in alcoholic beverages must be detected easily and reliably. However, the high similarity of methanol and ethanol makes methanol sensing surprisingly challenging.

Various methods to distinguish between methanol and ethanol have been reported, such as gas chromatography-mass spectrometry [1], spectrofluorometric [2], optical fiber sensing [3–5], high performance liquid chromatography (HPLC) [6], electrochemical methods [7], Raman spectroscopy [8], and multifunctional nanomaterials [9,10]. However, most of the above methods require cumbersome sample preparation, expensive equipment, and complicated operation, which greatly limits the practicality of the sensor. Therefore, a method allowing simple, convenient, and selective distinction between ethanol and methanol in aqueous solutions is urgently required.

Birefringent liquid crystals (LCs) are sensitive to the influence of environmental fields which change the effective refractive index of LC molecules and alter their optical properties. LC-based biosensors combine the advantages of high sensitivity, good repeatability, and good biocompatibility, so they have recently attracted considerable attention and are extensively used in physical sensing [11,12], chemical sensing [13,14], and biological sensing [15,16]. Cholesteric liquid crystals (CLCs) with natural periodicity can be regarded as one-dimensional photonic bandgap materials with central wavelength $\lambda_c = \bar{n}p$ and bandwidth $\Delta\lambda = \Delta np$, where $\bar{n}$ is average refractive index, Δn is the birefringence of the LC host, and p is CLC pitch length. The helical structure is susceptible to disturbances caused by external stimuli (e.g., temperature, electric voltage, chemical agents, pH environment), which in turn significantly change the CLC alignment and helical pitch, altering the wavelength of the reflected light [17]. This unique optical response, induced by external stimuli, can be transduced into macroscopic responses; for example, changes in the reflected colors of CLCs that are visible to the naked eye. Therefore, CLCs can act as stimuli-responsive materials for colorimetry, and their potential has been investigated for more and more chemical sensor applications over the last decade [18–26].

The unique property of CLCs has attracted great interest to develop visual sensing materials that use color change as a detection signal. To enhance the stability, repeatability, and ease of operation of CLC films, some researchers have proposed a sensor platform based on cholesteric liquid crystal polymer (CLCP) network structure, which can be fabricated easily on a single glass or plastic substrate [14,19,21–26]. Owing to the CLCP sensor can be implemented in a solid-state film with simple structure, high mechanical stability, convenient operation, and accurate measurement, it has become a hot research topic in the current sensing field [14,19,21–26]. Among these technologies of CLCP, C. K. Chang et al. proposed a new photonic CLC material, which is a hydrogen-bridged CLCP network with porosity, to distinguish methanol from ethanol [19,22]. To improve the expansion degree of the helical structure and enhance the sensing capability, they introduced porosity to the polymer networks. The creation of porosity in the CLCP networks not only facilitates the diffusion of the analytes into the CLC film, but also increases the surface-area-to-volume ratio of the polymer networks that react with the analytes. The hydrogen-bridged CLCP has carboxylic moieties, and its hydrogen bonds can be activated to interact with the hydroxyl group of analytes as treated with alkaline solution. As a result, the bond-breaking treatment stimulates the absorption of alcohol molecules having hydrogen bonds. These CLCP networks with the carboxyl groups having different affinities for methanol and ethanol molecules are very sensitive to alcohol molecules. When the carboxylate polymer absorbs alcohol molecules with different molecular polarities, different degrees of expansion occur, resulting in the CLC film showing different reflected wavelength changes.

Although the previous works have demonstrated the discrimination between methanol and ethanol by using hydrogen-bridged cholesteric polymer networks [19,22], the authors mainly discussed the sensing characteristics of CLCP films fabricated by different CLC mixtures to alcohol molecules. There are some problems that need to be resolved further, such as (1) the optimum activation time of alkaline solution treatment, (2) too long detection time (~10 min), (3) poor reversibility, and so on. In this work, the commercial diacrylate reactive mesogen LC242 was utilized as the host LC material to achieve a cost-effective biosensor for determining ethanol and methanol. LC242 cannot directly interact with alcohol molecules, so to differentiate methanol from ethanol, porous hydrogen-bonded CLC polymer (CLCP) networks were applied, in which the hydrogen bridges of benzoic acid based dimers in the networks were used and needed to be broken and activated using an alkaline solution. Here, to obtain the optimum activation time of CLCP networks for alkaline solution treatment, the transient spectra of CLCP film were monitored during the bond-breaking process. According to the experimental result, the activated time for breaking the hydrogen bond of CLCP was found to be ~130 s. In order to verify the ability of the prepared CLC film to detect alcohol solutions, the transmission spectra were measured using alcohol solutions with different methanol and ethanol mixing ratios. Alcohol absorption was monitored by measuring changes in the transmission band of the cholesteric network due to the increased helical pitch of the CLC when the polymer networks swelled by fast in-diffusion of alcohol molecules. Based on the experimental results, the total alcohol concentration and proportion of methanol and ethanol in solutions could be detected in 30 s by monitoring the reflected wavelength shift of CLCP. As the methanol–ethanol ratio decreased, the sensing sensitivity of the cholesteric networks to alcohol molecules increased.

Additionally, the influence of UV intensity for curing the CLC mixture on the sensing sensitivity was explored. Higher UV intensity was found to enhance sensitivity to alcohol molecules and improve discrimination between methanol and ethanol. Moreover, the thermal treatment at 50 °C for 1 min was also applied to make the alcohol molecules desorbed from the hydrogen-bridged CLC networks for recovering the CLCP films. It has been proved that CLCP networks still preserved good reversibility after five repeated tests. The response of CLCP to alcohols yielded quickly reversible color changes, indicating that hydrogen-bonded CLCP could quickly absorb and release alcohol molecules by thermal treatment. This work demonstrated that CLCP film had good specificity for the detection of alcohol solutions, provided rapid detection with high stability, and offered cost-effective and simple spectral or visual observation. As a first application, a CLCP network sensor for screening alcoholic beverages could help to prevent outbreaks of methanol poisoning, and it is believed that this technology has broad applicability and strong commercial development potential in the future. The CLCP sensor provides an easy to use, low-cost, rapid and repeatable method for the detection and quantification of methanol/ethanol contamination in solutions and shows strong potential as a cost-effective sensor for other liquid sensing applications.

2. Results and Discussion

To enable the absorption of alcohol, the hydrogen bonds between 4-(acryloyloxyhexyloxy) benzoic acid (6OBA) and 4-(6-acryloyloxy-n-hex-1-yloxy)-2-methylbenzoic acid (6OBAM), which are the hydrogen-bonding molecular triggers and form the dimer through the hydrogen bond, were broken and activated to form carboxylic salt ($COO^{-}Na^{+}$) using an alkaline solution, as shown in Figure 1. The duration of the bond-breaking process considerably affects the performance of the sensor, so to obtain the optimum bond-breaking process, the transient spectrum was recorded.

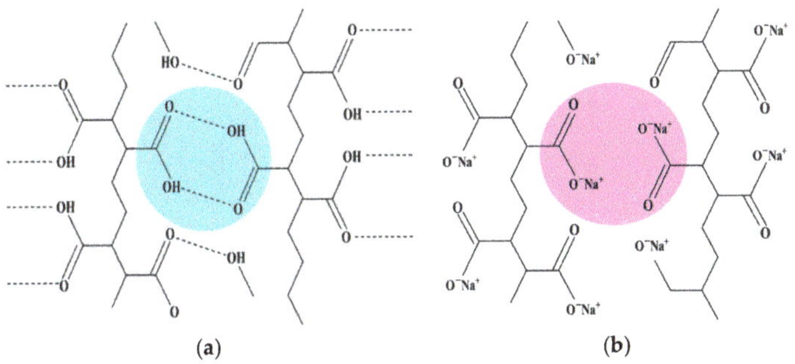

Figure 1. Schematic of (**a**) the dimer formed between 6OBA and 6OBAM through the hydrogen bonds; (**b**) the hydrogen bond was broken by NaOH, and then sodium carboxylate ($COO^{-}Na^{+}$) formed.

During the bond-breaking process, the reflected color of CLCP film cured at UV intensity of 20 mW/cm^2 was observed to gradually change from blue through purple to orange-red. The complete color change took 130 s, as shown in Figure 2a. The reflected color of CLCP film showed a red shift during the activation process, which was the result of swelling of the helical structure of the CLC due to the broken hydrogen bonds in the CLC networks and the accompanying absorption of alkaline solution. Figure 2b shows the time-dependent transmission spectrum of the CLCP film. The transmittance at the peak wavelength near 441 nm was observed to gradually increase with increasing immersion time, and another new transmittance band at peak wavelength near 565 nm was generated after a few seconds (~10 s). The transmittance of this new band decreased with immersion time, and its peak wavelength and bandwidth became progressively red-shifted and broader over time. However, when the immersion time was beyond 130 s, the original band at ~441 nm completely disappeared, and the new band had no obvious wavelength shift over time. Nevertheless, it was also found that too long soaking time can unexpectedly cause the CLCP film to detach from the substrate surface. According to the experimental results, the time required for breaking the hydrogen bond of CLCP was selected as 130 s in this work.

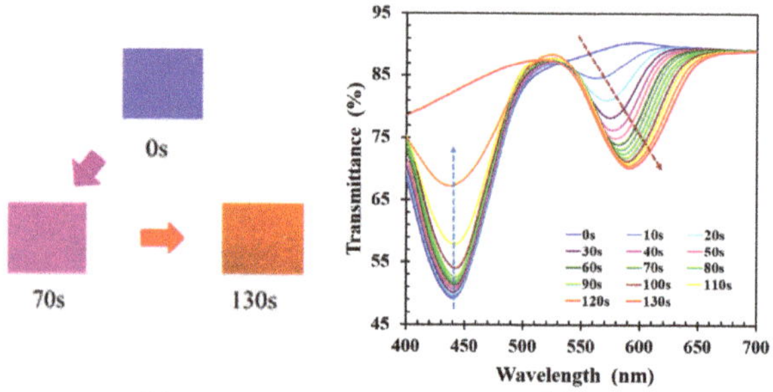

(a) (b)

Figure 2. (**a**) The change of the color reflected by CLC film, and (**b**) the transient transmission spectra of CLC film at different immersion times during the process of hydrogen bond breaking.

To study the alcohol sensing characteristics of the functionalized CLCP film, 40 μL of 40% ethanol solution was dropped onto the CLC film, and the corresponding dynamic spectral response of CLC was measured as shown in Figure 3a. It demonstrated that the alcohol solution caused the wavelength to shift from green to red within 5 s, and the new transmission band was broader than that of the original CLCP film. The new peak wavelength of the CLC film red-shifted further with time as the number of ethanol molecules absorbed by the CLC film increased. The red-shift phenomenon was due to rapid absorption of ethanol molecules by carboxylic salt groups, resulting in expansion of the polymer networks due to swelling, and a decrease of the order of the CLC helical structure which caused the transmission band to widen. Maximum red shift occurred after ~20 s and was followed by a small blue shift after around 30 s. This unexpected blue shift, with a small accompanying decrease in bandwidth, is highlighted by the blue arrow in Figure 3a and shown in detail in Figure 3b. The transmission band for blue shift was narrower than that for red shift, and the amount of blue shift was minor. The minor blue shift was likely due to the release or desorption from the film of a small fraction of alcohol molecules, leading to shrinkage. Shrinkage of the film resulted in a decrease of the pitch, causing a blue shift of the transmission band and an increase in the order of the CLC helical structure.

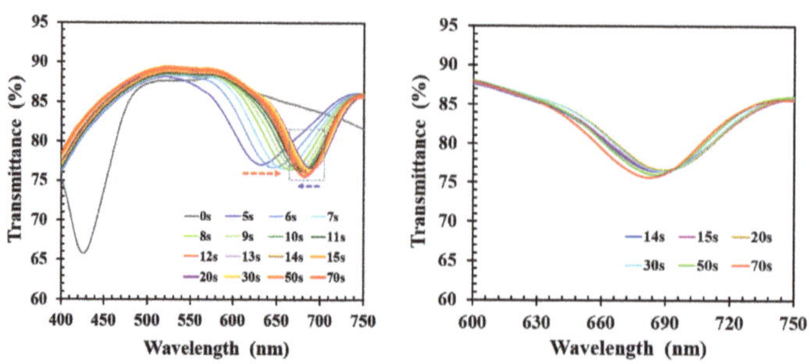

(a) (b)

Figure 3. (**a**) Time lapse transmittance spectra of CLC film subjected to 40% ethanol solution, and (**b**) scale up of the blue-shifted region indicated by the blue arrow in (**a**).

Figure 4 shows the time-dependent wavelength shift of the CLC film under different ethanol concentrations. The experimental results showed that the wavelength shift increased rapidly with detection time and reached a maximum at a certain time, as well as more ethanol molecules contained in the alcohol solution caused the CLC film to exhibit a larger red shift. Nonetheless, the wavelength shift decreased slightly with time after the time of wavelength shift reached the maximum value. The tiny decrease at longer times may be because there were no more ethanol molecules interacting with the hydrogen-bridged CLC networks and at the same time, the ethanol molecules in the CLC film were desorbing, so a small amount of wavelength blue shift occurred. These results showed that CLC films were able to quickly produce a visible response that was sensitive to ethanol concentration.

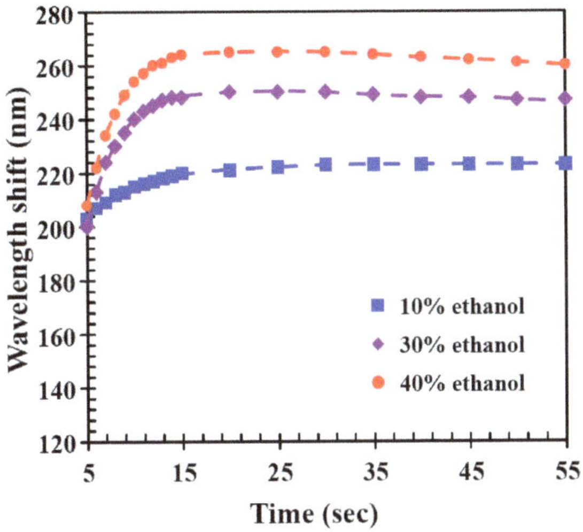

Figure 4. Time-dependent wavelength shifts of CLC film upon exposure to solutions with different ethanol concentrations.

To evaluate the ability of hydrogen-bonded CLCP networks to sense and differentiate ethanol and methanol, solutions were prepared with 60% water and 40% methanol–ethanol mixture, in which the methanol–ethanol ratio was varied. Forty microliters of alcohol solution was dropped onto the CLCP film, and the transmission spectrum was measured 30 s later. Figure 5a shows the spectral response of the CLCP film to the various alcohol solutions. It can be seen the wavelength shifts induced by the uptake of methanol and ethanol molecules were significantly different. This was due to the different molecular affinities of methanol and ethanol with hydrogen-bonded CLC. The molecular affinity of ethanol is larger than that of methanol with hydrogen-bonded CLC polymer networks [19,22], so alcohol solutions containing more ethanol molecules cause the helical structure of the CLC to swell more. Therefore, the central wavelength of the CLCP film showed a larger red shift as the ratio of methanol–ethanol was reduced. Figure 5b shows the wavelength shift of the CLC films with different methanol–ethanol ratios, in which a linear fit between wavelength shift and the alcohol concentration and the ratio of methanol/ethanol was obtained by the least squares method. The slope of the linear fit denotes the selectivity of the CLC sensor between methanol and ethanol for a fixed total alcohol concentration of 40%.

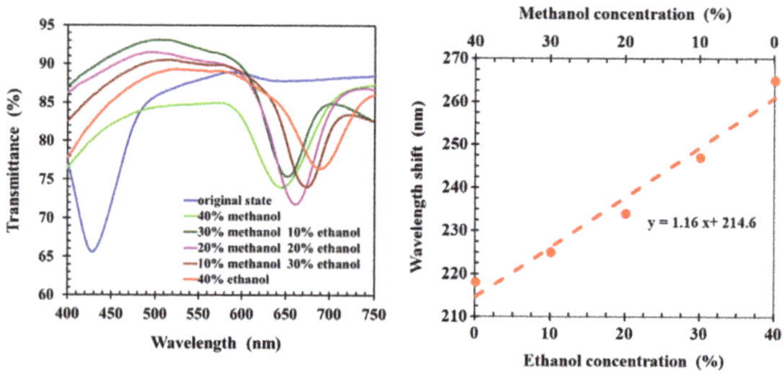

(a) (b)

Figure 5. (**a**) The transmission spectra of CLC films, (**b**) wavelength shift of CLCs induced by alcohol solutions (40%) with different ethanol–methanol ratios.

The sensing performance of the CLCP film was determined at different alcohol concentrations from 0% to 40% and different ethanol–methanol ratios. Figure 6 shows that wavelength red shift increased with increasing the alcohol concentration, regardless of the mixture of methanol and ethanol at any ratio. At the same alcohol concentration, the greater the proportion of ethanol, the greater the wavelength shift. This phenomenon was attributed to the molecular affinity of ethanol with a hydrogen-bonded CLCP networks higher than that of methanol, so the contribution of the same amount of ethanol to the wavelength shift was larger than that of methanol. Linear fits are shown for wavelength shift versus total alcohol concentration, and the sensing sensitivity of CLC networks to alcohol molecules increased as the methanol–ethanol ratio declined. The sensitivity for ethanol and methanol was 1.35 and 0.18 nm/%, respectively. Consequently, the CLCP films illustrated good selective responses to methanol from ethanol. Additionally, from the experimental results of Figure 6, it could be also observed the transmission central wavelength of the CLC film immersed in pure water exhibited a red shift to around 210 nm, which was ~48.8% wavelength-increment caused by shrinkage of the helical pitch. The swelling degree of the porous film in alcohol solutions mixed by altered ratio of methanol and ethanol was considerably different, for example, the expansion degree of helical structure induced by 40% ethanol and methanol solution was obtained as ~61.4% and ~50.5%, respectively. Compared with the previous works [19,22], in which the transmission spectrum was recorded after soaking the sample in solutions for 10 min, although the sensitivity of our CLCP films for sensing alcohol molecule was lower, a very small volume (~40 μL) of alcohol solution could be distinguished in a very short time. The detection time (~30 s) was significantly shortened.

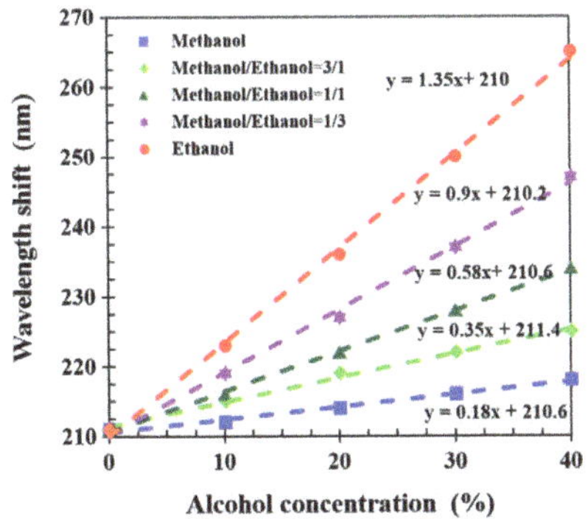

Figure 6. The wavelength shift of CLCP films induced by alcohol solutions, with different total alcohol concentrations and methanol–ethanol ratios.

The UV curing intensity during the polymerization process affects the size of the voids within the polymer networks [27,28], significantly influencing the response of CLC films to alcohol solutions. Higher curing intensity generally results in more open polymer networks with larger voids. High porosity CLC polymer networks can increase the diffusion rate of alcohol molecules and the expansion of the helical structure of CLC, thereby enhancing the sensitivity to analytes. Therefore, in addition to using nonreactive mesogen 5CB as template molecules to create the porous structure, the microstructure of the CLC polymer networks could also be controlled by the photopolymerization conditions to improve the sensing performance.

The standard UV curing intensity used in this work was 20 mW/cm^2. To study the impact of UV curing intensity on the sensing performance of the CLC film, a reduced intensity of 5 mW/cm^2 was applied to cure the CLC film, and the transient spectral response to 40% ethanol solution was measured as shown in Figure 7a. Comparison with the 20 mW/cm^2 intensity results presented in Figure 3a shows that reduced UV intensity gave very similar overall behavior. To examine the differences, Figure 7b plots the time-dependent wavelength shifts of CLC films realized with UV intensities of 20 and 5 mW/cm^2. Both the ethanol-induced wavelength shifts rapidly increased with detection time, reaching a maximum value at 20 s and slowly decaying 30 s later. The time-dependent wavelength shift was larger for higher UV cure intensity because the structure of CLC polymer networks realized by higher UV intensity was relatively sparse and had larger voids. This was because a rougher CLC structural morphology with larger voids allowed the alcohol molecules to easily diffuse into the CLC film and promoted a more significant expansion of the CLCP pitch structure, so that a greater wavelength shift response was induced.

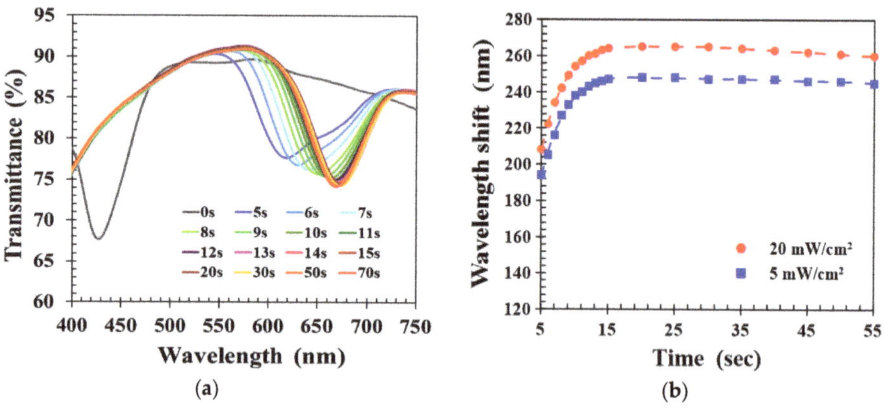

Figure 7. (**a**) Time lapse transmission spectra of CLC film fabricated by UV intensity of 5 mW/cm^2, and (**b**) comparison of time-dependent wavelength shifts of CLC films realized by different UV intensities.

Figure 8 shows the dependence of the wavelength shift of CLCP films realized with UV curing intensities of 5 and 20 mW/cm^2 on ethanol concentration, in which the spectral response was measured at a detection time of 30 s. Linear fits gave sensitivities of 0.95 and 1.35 nm/% for CLCP films cured at intensities of 5 and 20 mW/cm^2, respectively. It can be observed that the CLC film cured by higher UV intensity had a higher sensitivity, and higher UV intensity could shorten the process time to cure the CLC film. It concludes that sensor sensitivity could be enhanced by employing an optimum CLC network microstructure to facilitate the diffusion of alcohol molecules into the film and the expansion of the helical structure.

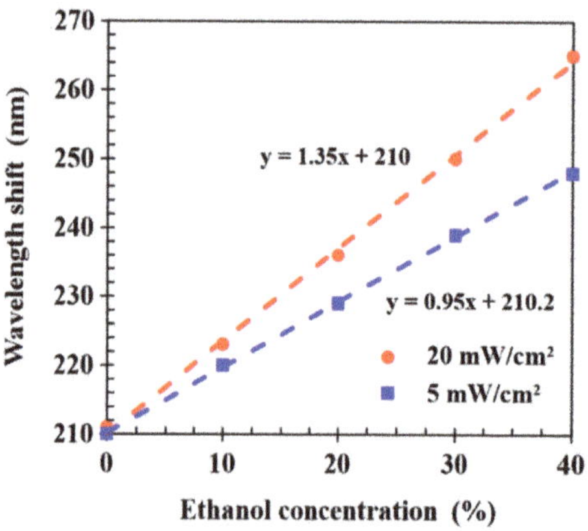

Figure 8. Wavelength shifts of CLCP films cured by different UV intensities and exposed to ethanol concentrations from 0 to 40%.

In some applications, it is desirable for a sensor to be reusable. To study the reversibility and repeatability of the spectral response to analytes, the thermal treatment was applied to detach the alcohol molecules from CLCP networks. First, the CLCP film was immersed in 15 mL of 40% ethanol solution for ~30 s. Then, to desorb the alcohol molecules, the CLCP film was taken out of the solution and baked at 50 °C for 1 min after blowing off any liquid droplets on the film surface with a nitrogen stream. Repeating the above immersion and baking process, the individual spectrum of the CLC film after baking was measured, as shown in Figure 9. It can be observed that the helical pitch of the CLCP film could nearly be restored to its original state without any significant loss in performance during these five cycles, indicating the response of CLCs to alcohol yields reversible and repeatable optical response changes. The reversible response was presumably due to the release of alcohol molecules from the CLCP film, leading to restoration of the helical structure. Based on the experimental results, CLCP film could rapidly adsorb and desorb alcohol molecules, which demonstrates that a real-time sensor for discriminating ethanol and methanol in aqueous solution in a fast and reversible manner could be constructed.

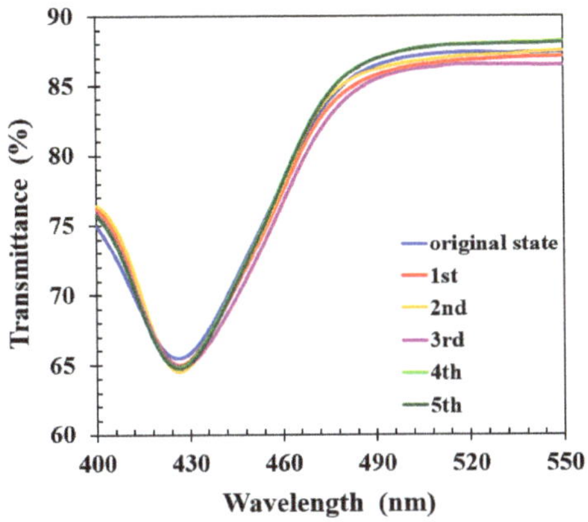

Figure 9. The transmittance spectra of CLC after repeated cycles of immersion in alcohol solution followed by baking.

The extraordinary property of CLCP can be applied to develop visual sensing materials to alcohol molecules using color changes as a detection signal. It provides a new strategy for measuring the concentration of methanol and ethanol in solution. Moreover, CLCP film with high mechanical stability can be simply fabricated by depositing the LC mixture onto a substrate by inkjet printing or doctor blade coating, potentially enabling the fabrication of large areas in roll-to-roll (R2R) processing at low cost. Therefore, the CLCP film is an excellent candidate for the responsive alcohol sensor, due to its characteristics of ease of manufacture, visibility to the naked eye, fast response, high stability, label-free, battery-free and a microliter volume requirement for the bio-sample solution.

3. Materials and Methods

3.1. Preparation of CLCP Film

To preserve the polymer integrity of the CLC film, the reactive mesogen and chiral dopant used in this study were diacrylate LC242 and LC756, respectively, obtained from BASF Co. Ltd. (Ludwigshafen, Germany) Paliocolor LC242 is a non-alcohol-responsive reactive mesogen. To facilitate responding alcohol molecules, the photopolymerizable alkyloxybenzoic acids 6OBA and 6OBAM (Synthon Chemicals Ind. Ltd., Bitterfeld-Wolfen, Germany) were used. Additionally, to obtain a fast response and enhance the sensitivity of the CLCP networks on the uptake of the analytes, a nonreactive mesogen 5CB (Tokyo Chemical Ind. Ltd., Tokyo, Japan) as a porogen was also added in the CLC mixture to create macroscopic pores in the polymer networks [19,22]. The free radical photoinitiator used was 2,2-dimethopxy-1,2-diphenyl-ethanone (IRG651, CIBA Co. Ltd., Ikeja, Nigeria). In order to make the transmittance spectra of CLCP upon exposure to alcohol solution cover the visible range, the CLCP film was made by formulating a chiral liquid-crystal mixture of LC242, LC756, 5CB, 6OBA, 6OBAM, and IRG651 in a weight ratio of 34.2:4.8:18:21:21:1. All the LC mixtures were dissolved in a tetrahydrofuran (THF) solution (4:6).

To fabricate the CLCP specimen, a polyimide alignment layer (SE-3140, Nissan Chemical Ind. Ltd., Tokyo, Japan) was first coated on a 2.5 $\times$ 2.5 cm cleaned glass substrate, cured, and then rubbed with a velvet cloth to induce planar alignment. Forty microliters of CLC mixture was dropped onto the rubbed substrate, heated at 62 °C for 15 min to volatilize the THF solvent, and then a cover glass substrate coated with fluorinated alkyl silane was attached. After cooling the sample to 59 °C, a shearing force was applied in a horizontal direction to obtain a well-aligned CLC film. Subsequently, the sample was photopolymerized by UV irradiation at intensity 20 mW/cm^2 for 10 min, and the cover glass was detached after completing the photopolymerization process. To generate the porosity required to enhance sensing ability, the CLCP film was heated on a hot plate at 105 °C for 5 h to extract the nonreactive mesogen 5CB after the polymer networks have been formed. Figure 10 shows the scanning electron micrograph (SEM) of the porosity created in the CLC polymer networks after the extraction of 5CB by heat treatment. The size and density of the porosity in the CLC polymer networks can be controlled by the amount of 5CB in the mixture and the photopolymerization condition [28].

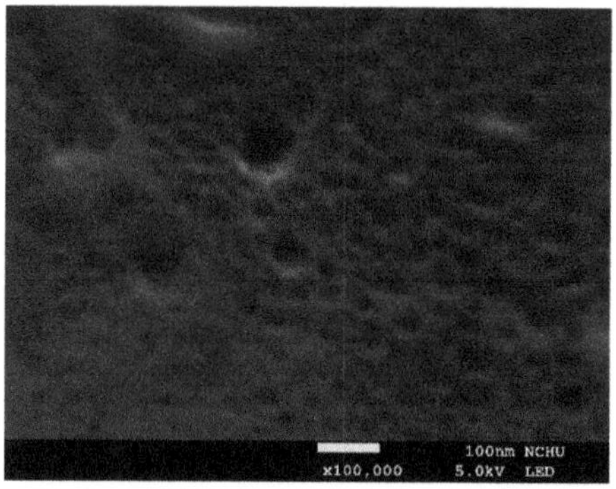

Figure 10. The SEM image is a cross-sectional micrograph of the CLCP film after the extraction of 5CB.

Figure 11 shows the transmission spectra of the CLCP film at different stages. The transmission central wavelength and reflected color of the CLCP would reveal its helical pitch. Initially, a CLCP film was obtained with central wavelength around 536 nm. Upon extraction of 5CB by heat treatment, the transmission central wavelength of the CLC film exhibited a blue shift to around 441 nm (17.7% wavelength reduction) caused by shrinkage of the helical pitch. The reduction ratio in the central wavelength of the CLC film was close to the ratio for the nonreactive mesogen 5CB doped in the CLC mixture, confirming that the heat treatment effectively removed 5CB from the CLCP film.

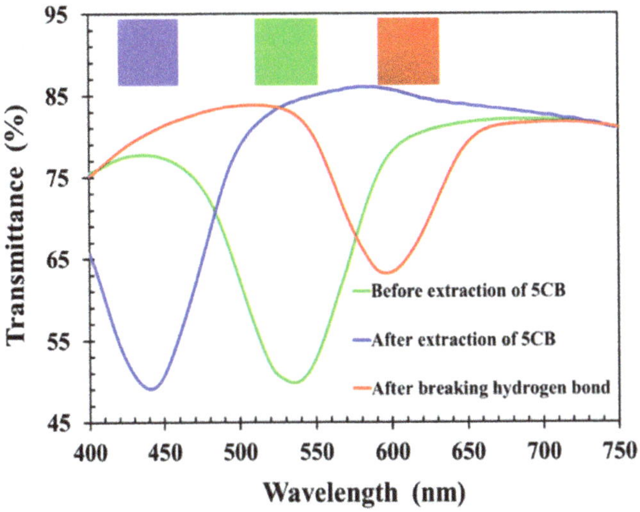

Figure 11. Transmission spectra of CLC films at different process steps.

The photopolymerized alkyloxybenzoic acids of 6OBA and 6OBAM stuck to each other physically to constitute a hydrogen bond by means of the carboxylic moieties [19,22,23]. The carboxylic moieties of CLCP activated during breaking of the hydrogen bridges interacted with the hydroxyl group of analytes, so the bond-breaking treatment stimulated the absorption of alcohol molecules having hydrogen bonds. To enable the absorption of alcohol molecules, the process of breaking hydrogen bonds was performed after the extraction of 5CB by soaking the CLCP film in 20 mL of sodium hydroxide solution (0.5 M). During the process of breaking hydrogen bridges, 6OBA and 6OBAM form carboxylic salt ($COO^{-}Na^{+}$), and the helical structure of CLCP with broken hydrogen bond swells. As shown in Figure 11, the spectrum of the CLCP film shows a red shift to around 598 nm after the activation in the alkaline solution. This was a consequence of the CLCP helical structure swelling due to the broken hydrogen bonds in the CLCP networks and the accompanying absorption of the alkaline buffer solution. However, when the CLCP salt film was dried on a hot plate at 50 °C for 3 min, the CLCP film showed a blue shift and returned to the original reflected color after the drying process.

Furthermore, the diffusion rate of alcohol into the CLCP film is critically impacted by the morphology of the polymer network microstructure, which is significantly influenced by the UV curing conditions such as UV light intensity [27,28]. To further address the influence of the microstructure of the CLCP networks on the sensing characteristics of CLC film to alcohol molecules, the UV light at a lower intensity of 5 mW/cm^2 was also applied to construct the CLCP film to study the ability to differentiate alcohol solutions.

3.2. *Characterization of the CLCP-Based Responsive Alcohol Sensor*

To assess the alcohol-responsive ability of the CLCP film, the transmission spectra of CLCP films were measured for different alcohol solutions. Figure 12 shows a schematic diagram of the test system. A halogen lamp and a Spectra Academy SV-2100 spectrophotometer were used as the incident light source and for measuring the optical spectrum of the CLCP, respectively. To quantitatively investigate the impact of alcohol concentration on the response characteristics of CLCP, 40 μL alcohol solutions with different alcohol concentrations and ratios of methanol to ethanol were dropped onto dried CLC polymer salt films, and the transient spectral characteristics of the CLCP were measured in real time.

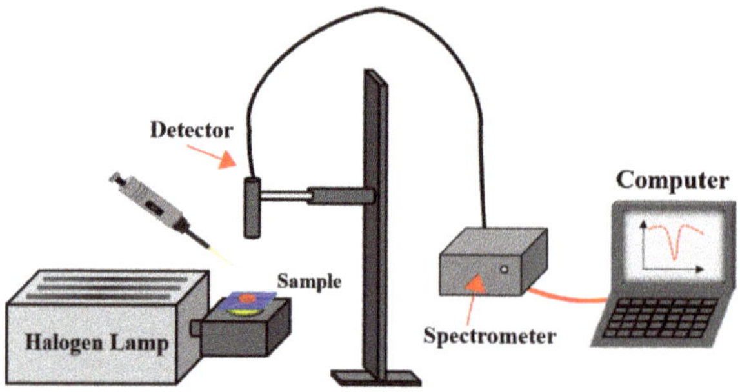

Figure 12. Schematic diagram of the system for evaluating the sensing properties of CLC film.

4. Conclusions

The hydrogen-bonded CLCP film could realize rapid detection of alcohol solutions containing mixed ethanol and methanol and could distinguish methanol and ethanol. The experimental results revealed that ethanol and methanol molecules could make the reflected color of CLCs red shift, while ethanol caused a larger red shift. The helical structure of CLCs swelled upon the adsorption of alcohol molecules, causing a red shift of the transmission band in under 30 s. A highly linear relationship was found between wavelength shift and alcohol concentration. The sensing sensitivity of CLCP networks to alcohol molecules increased as the alcohol concentration increased and the methanol–ethanol ratio decreased. The sensitivities of CLCP networks were 1.35 and 0.18 nm/% to ethanol and methanol, respectively. It was also demonstrated that CLCP film cured with high light intensity had a better optical response to alcohol solution than that cured by low light intensity. Furthermore, it was verified that CLCP film could rapidly absorb and release alcohol, leading to reversible and repeatable optical response changes appropriate for reusable sensor designs. As a result, the CLCP film offers the basis for stable, fast, sensitive, and reversible sensors, which are highly selective due to the differentiated adsorption of alcohols into the micropores of the CLC polymer salt. It is believed that CLCP film holds great potential for use in biosensor applications because of its advantages of linear response and visible color.

Author Contributions: Conceptualization, S.-J.H. and M.-F.L.; methodology, T.-Y.Y., M.-F.L. and R.-D.L.; validation, S.-J.H. and T.-Y.Y.; formal analysis, S.-J.H. and T.-Y.Y.; resources, S.-J.H.; writing—original draft preparation, T.-Y.Y.; writing—review and editing, S.-J.H. All authors have read and agreed to the published version of the manuscript.

Funding: This research was funded by the Ministry of Science and Technology of Taiwan (MOST 108-2221-E-239-022-).

Institutional Review Board Statement: Not applicable.

Informed Consent Statement: Not applicable.

Data Availability Statement: No new data were created or analyzed in this study. Data sharing is not applicable to this article.

Acknowledgments: The authors also sincerely acknowledge BASF Taiwan Ltd. for providing the reactive mesogen and chiral dopant monomer materials.

Conflicts of Interest: The authors declare no conflict of interest.

Sample Availability: Samples of the CLCP are available from the authors.

References

1. Tiscione, N.B.; Alford, I.; Yeatman, D.T.; Shan, X. Ethanol Analysis by Headspace Gas Chromatography with Simultaneous Flame-Ionization and Mass Spectrometry Detection. *J. Anal. Toxicol.* **2011**, *35*, 501–511. [CrossRef] [PubMed]

2. Solís, R.A.; González, J.V.; Brenes, J.P.S.; Hernández, G.E. Design of a spectrofluorometric method for the determination of ethanol in air. *Uniciencia* **2013**, *27*, 215–231.
3. Su, Y.; Lan, Z.; Wang, J.; Zeng, L.; Zhou, D.; Peng, Z.; Sun, W.; Liu, Y. Optical Fiber Sensor for Determination of Methanol Ratio in Methanol-Doped Ethanol Based on Two Cholesteric Liquid Crystal Droplets Embedded in Chitosan. *J. Lightwave Technol.* **2021**, *39*, 5170–5176. [CrossRef]
4. Iqbal, F.; Biswas, S.; Bulbul, A.A.-M.; Rahaman, H.; Hossainc, M.B.; Rahaman, M.E.; Awal, M.A. Alcohol sensing and classification using PCF-based sensor. *Sens. Bio-Sens. Res.* **2020**, *30*, 100384. [CrossRef]
5. Liu, D.; Kumar, R.; Wei, F.; Han, W.; Mallik, A.K.; Yuan, J.; Wan, S.; He, X.; Kang, Z.; Li, F.; et al. High sensitivity optical fiber sensors for simultaneous measurement of methanol and ethanol. *Sens. Actuators B* **2018**, *271*, 1–8. [CrossRef]
6. Chen, S.H.; Wu, H.L.; Yen, C.H.; Wu, S.M.; Lin, S.J.; Kou, H.S. Trace determination of methanol in water-ethanol solution by derivatization and high-performance liquid chromatography. *J. Chromatogr. A* **1998**, *799*, 93–99. [CrossRef]
7. Hazra, A.; Bhowmik, B.; Dutta, K.; Chattopadhyay, P.P.; Bhattacharyya, P. Stoichiometry, Length, and Wall Thickness Optimization of TiO_2 Nanotube Array for Efficient Alcohol Sensing. *ACS Appl. Mater. Interfaces* **2015**, *7*, 9336–9348. [CrossRef]
8. Xu, Q.; Ye, Q.; Cai, H.; Qu, R. Determination of methanol ratio in methanol-doped biogasoline by a fiber Raman sensing system. *Sens. Actuators B* **2010**, *146*, 75–78. [CrossRef]
9. Ishihara, S.; Iyi, N.; Labuta, J.; Deguchi, K.; Ohki, S.; Tansho, M.; Shimizu, T.; Yamauchi, Y.; Sahoo, P.; Naito, M.; et al. Naked-Eye Discrimination of Methanol from Ethanol Using Composite Film of Oxoporphyrinogen and Layered Double Hydroxide. *ACS Appl. Mater. Interfaces* **2013**, *5*, 5927–5930. [CrossRef]
10. Chen, Y.; Zeng, X.; Jiang, G.W.; Wei, D.; Ling, M.; Ng, K.W.; Qin, Y. Ultrasensitive ethanol sensor based on segregated zno- In_2O_3 porous nanosheets. *Appl. Surf. Sci.* **2021**, *535*, 147697.
11. Hu, J.Y.; Zhou, D.; Su, Y.; Liu, S.; Miao, P.; Shi, Y.; Sun, W.; Liu, Y. Fiber micro-tip temperature sensor based on cholesteric liquid crystal. *Opt. Lett.* **2020**, *45*, 5209–5212. [CrossRef] [PubMed]
12. Zhang, M.Z.; Ge, Y.M.; Wu, S.T.; Hwang, S.J. Ultra-sensitive curvature sensor based on liquid crystal-infiltrated fiber interferometer. *J. Phys. D Appl. Phys.* **2017**, *50*, 015102. [CrossRef]
13. Huang, J.W.; Chang, J.J.; Yang, R.H.; Chen, C.H. Agarose dispersed liquid crystals as a soft sensing platform for detecting mercuric ions in water. *Res. Chem. Intermediat.* **2019**, *45*, 5409–5423. [CrossRef]
14. Wang, T.H.; Liu, M.F.; Hwang, S.J. Optical sensing of organic vapour based on polymer cholesteric liquid crystal film. *Liq. Cryst.* **2020**, *47*, 1390–1397. [CrossRef]
15. Lin, C.T.; Hsu, W.T.; Hwang, S.J. Real-time liquid crystal-based creatinine sensor using a micro-patterned flexible substrate. *Liq. Cryst.* **2021**, *48*, 1660–1670. [CrossRef]
16. Lee, M.J.; Pai, C.P.; Wu, P.C.; Lee, W. Label-free single-substrate quantitative protein assay based on optical characteristics of cholesteric liquid crystals. *J. Mol. Liq.* **2021**, *331*, 115756. [CrossRef]
17. Mitov, M. Cholesteric liquid crystals with a broad light reflection band. *Adv. Mater.* **2012**, *24*, 6260–6276. [CrossRef]
18. Tang, J.Y.; Fang, J.B.; Liang, Y.L.; Zhang, B.; Luo, Y.H.; Liu, X.Y.; Li, Z.B.; Cai, X.J.; Xian, J.Q.; Lin, H.; et al. All-fiber-optic VOC gas sensor based on side-polished fiber wavelength selectively coupled with cholesteric liquid crystal film. *Sens. Actuators B* **2018**, *273*, 1816–1826. [CrossRef]
19. Chang, C.K.; Bastiaansen, C.W.; Broer, D.J.; Kuo, H.L. Discrimination of alcohol molecules using hydrogen-bridged cholesteric polymer networks. *Macromolecules* **2012**, *45*, 4550–4555. [CrossRef]
20. Zhang, P.; Kragt, A.J.J.; Schenning, A.P.H.J.; Haan, L.T.; Zhou, G. An easily coatable temperature responsive cholesteric liquid crystal oligomer for making structural colour patterns. *J. Mater. Chem. C* **2018**, *6*, 7184–7187. [CrossRef]
21. Yang, Y.; Zhou, D.; Liu, X.; Liu, Y.; Liu, S.; Miao, P.; Shi, Y.; Sun, W. Optical fiber sensor based on a cholesteric liquid crystal film for mixed VOC sensing. *Opt. Express* **2020**, *28*, 31872–31881. [CrossRef] [PubMed]
22. Chang, C.K.; Bastiaansen, C.M.W.; Broer, D.J.; Kuo, H.L. Alcohol-Responsive, Hydrogen-Bonded, Cholesteric Liquid-Crystal Networks. *Adv. Funct. Mater.* **2012**, *22*, 2855–2859. [CrossRef]
23. Moirangthem, M.; Arts, R.; Merkx, M.; Schenning, A.P. An Optical Sensor Based on a Photonic Polymer Film to Detect Calcium in Serum. *Adv. Funct. Mater.* **2016**, *26*, 1154–1160. [CrossRef]
24. Lugger, S.J.D.; Houben, S.J.A.; Foelen, Y.; Debije, M.G.; Schenning, A.P.H.J.; Mulder, D.J. Hydrogen-Bonded Supramolecular Liquid Crystal Polymers: Smart Materials with Stimuli-Responsive, Self-Healing, and Recyclable Properties. *Chem. Rev.* **2021**. Published Online. [CrossRef]
25. Liu, B.; Yang, T.; Mu, X.; Mai, Z.; Li, H.; Wang, Y.; Zhou, G. Smart Supramolecular Self-Assembled Nanosystem: Stimulus-Responsive Hydrogen-Bonded Liquid Crystals. *Nanomaterials* **2021**, *11*, 448. [CrossRef]
26. Herzer, N.; Guneysu, H.; Davies, D.J.D.; Yildirim, D.; Vaccaro, A.R.; Broer, D.J.; Bastiaansen, C.W.M.; Schenning, A.P.H.J. Printable Optical Sensors Based on H-Bonded Supramolecular Cholesteric Liquid Crystal Networks. *J. Am. Chem. Soc.* **2012**, *134*, 7608–7611. [CrossRef]
27. Dierking, I.; Kosbar, L.L.; Lowe, A.C.; Hel, G.A. Polymer network structure and electro-optic performance of polymer stabilized cholesteric textures II. The effect of UV curing conditions. *Liq. Cryst.* **1998**, *24*, 397–406. [CrossRef]
28. Li, W.; Cao, H.; Kashima, M.; Liu, F.; Cheng, Z.; Yang, Z.; Zhu, S.; Yan, H. Control of the Microstructure of Polymer Network and Effects of the Microstructures on Light Scattering Properties of UV-Cured Polymer-Dispersed Liquid Crystal Films. *J. Polym. Sci. Part B Polym. Phys.* **2008**, *46*, 2090–2099. [CrossRef]

molecules

MDPI

Article

Three Rings Schiff Base Ester Liquid Crystals: Experimental and Computational Approaches of Mesogenic Core Orientation Effect, Heterocycle Impact

Shady Nada [1], Mohamed Hagar [1,*], Omaima Farahat [1], Ahmed A. Hasanein [1], Abdul-Hamid Emwas [2], Abeer Ali Sharfalddin [3], Mariusz Jaremko [4] and Mohamed A. Zakaria [1]

1 Department of Chemistry, Faculty of Science, Alexandria University, Alexandria 21321, Egypt; shadyadel4869@gmail.com (S.N.); oom_farahat@yahoo.com (O.F.); ahmedhasanein@alexu.edu.eg (A.A.H.); mohamed.zakaria@alexu.edu.eg (M.A.Z.)
2 Core Labs, King Abdullah University of Science and Technology, P.O. Box 4700, Thuwal 23955-6900, Saudi Arabia; abdelhamid.emwas@kaust.edu.sa
3 Department of Chemistry, Faculty of Science, King Abdulaziz University, P.O. Box 80203, Jeddah 21589, Saudi Arabia; sharfalddin.aa@hotmail.com
4 Smart-Health Initiative (SHI) and Red Sea Research Center (RSRC), Division of Biological and Environmental Sciences and Engineering (BESE), King Abdullah University of Science and Technology (KAUST), P.O. Box 4700, Thuwal 23955-6900, Saudi Arabia; mariusz.jaremko@kaust.edu.sa
* Correspondence: mohamed.hagar@alexu.edu.eg

Abstract: Three rings 2-hydroxypyridine liquid crystalline compounds have been prepared and fully characterized. The mesomorphic behavior of the prepared compounds has been investigated in terms of differential scanning calorimetry (DSC) and polarized optical microscopy (POM). Moreover, a comparative study between the prepared compounds and previously reported analogs has been discussed in terms of the orientation and position of the mesogenic core, in addition to the direction of the terminal alkyl chains. Furthermore, a detailed computational approach has been studied to illustrate the effect of geometrical and dimensional parameters on the type of the enhanced texture and the mesomorphic range and stability. The results of the DFT study revealed that the orientation of the mesogen could affect the mesomorphic behavior and this has been attributed in terms of the degree of the polarizability of the linking groups. This result has been confirmed by calculation of the net dipole moment and the molecular electrostatic potential that show how the mesogen orientation and position could impact the molecular charge separation. Finally, the effect of the pyridyl group has been also investigated in terms of the calculated aromaticity index and the π-π stacking.

Keywords: Schiff base ester liquid crystals; DFT; heterocycles impact

1. Introduction

Liquid crystals (LCs) represent a state of matter that is thermodynamically placed between the solid phase and liquid phase [1]. Commonly, LC is called a mesomorphic state [2] thus, they have combined properties of them yet expressed special electro-optic phenomena which are unmatched in crystals or liquids. A usual liquid crystal molecule is represented by two main components: a mesogen which is a central rigid part and a spacer which represents the flexible side chains. Furthermore, the rod-like structure consists of a central linkage part between two ring systems [3]. Meanwhile, the physical properties are influenced by the orientation of the anisotropic molecules. Changing the orientation could cause altering in the mechanical properties of the medium and its optical characteristics. This is the main factor that many devices that utilized liquid crystal are based on, such as display panels [4,5]. Moreover, LCs are used recently in optical sensors for imaging trypsin activity [6], organic photovoltaic cells, transistors that benefit from the self-assembly feature [7,8], light polarization [9] and artificial muscles that are employing elastomers, which were formed by weakly cross-linking thermotropic liquid crystal phase with reactive mesogens [10].

Recently, Schiff base LCs widely attracted the attention of the LC community because of the feasibility and low cost of synthetic procedures. They are generally synthesized through the condensation reaction between carbonyl compounds and primary amines to form the (-CH=N-) imine linking group connecting the rigid core components [11]. Additionally, they could offer higher stability and allow mesophase formation [12] which would lead to interesting results such as the formation of room temperature ferroelectric liquid crystals [13], solar energy applications [14], and high efficient flame-retardation and anti-dripping action for polyesters [15]. The ester and imine linking units are the basic structural components for producing the mesomorphism of three aromatic rings thermotropic liquid crystals [16].

In our previous work, we had studied the impact of different lateral substituents [17] and the effect of the proportion of dialkoxy chain length [18] on the mesophase behaviour. We also synthesized new natural fatty acids and observed that mesophases depend on the length of the terminal alkenyl fatty acid chains and were highly impacted by their conformation [19]. Moreover, we had found that Schiff base compounds that contain two rings were mesomorphic except the NO_2 and the unsubstituted derivatives [20] and we synthesized four rings of Schiff bases/esters and extensively studied their mesophase behavior [21]. It was noticed that the nature of terminal side chains affected the mesomorphic properties of thermotropic calamitic liquid crystals [22].

In order to continue our study on Schiff bases with esters, here in the present work we will investigate the type and orientation of the mesogens and study their effect upon the mesophase stability in addition to the effect of the orientation of terminal groups. The investigation will include a computational DFT study. We will do a comparative study between newly synthesized compounds labelled **I C_8** and **I C_{16}** and compounds from our previous works labelled **II C_6** [23], **III C_6** [24], **IV C_8** [25], and **V C_6** [26]. Figure 1 illustrates the structures of the investigated compounds.

I C_8; X = N, n = 8
I C_{16}; X = N, n = 16
II C_6; X = C, n = 6

III C_6; X = H, Y = OC_6H_{13}, Z = N, L = C
IV C_8; X = OC_8H_{17}, Y = H, Z = N, L = C
V C_6; X = OC_6H_{13}, Y = H, Z = C, L = N

Figure 1. The structures of investigated compounds **I C_8**, **I C_{16}**, **II C_6**, **III C_6**, **IV C_8**, and **V C_6**.

2. Materials and Methods

2.1. Materials

N,N′-dicyclohexylcarbodiimide (DCC) (99%), 4-dimethylaminopyridine (DMAP) (98%), ethanol (95%), hydrochloric acid (38%), dichloromethane (99.8%), and 2-hydroxypyridine (98%) were purchased from Sigma Aldrich (Hamburg, Germany). 4-formylbenzonitrile (95%) and 4-octyloxyaniline (98%) were purchased from Aldrich (St. Louis, MO, USA). 4-n-hexadecyloxyaniline (98%) was purchased from Alfa Aesar (Heyvril, MA, USA). All chemicals were used without further purification.

2.2. *Synthesis*

The compounds **I** $C_{(8, 16)}$ were synthesized according to Scheme 1.

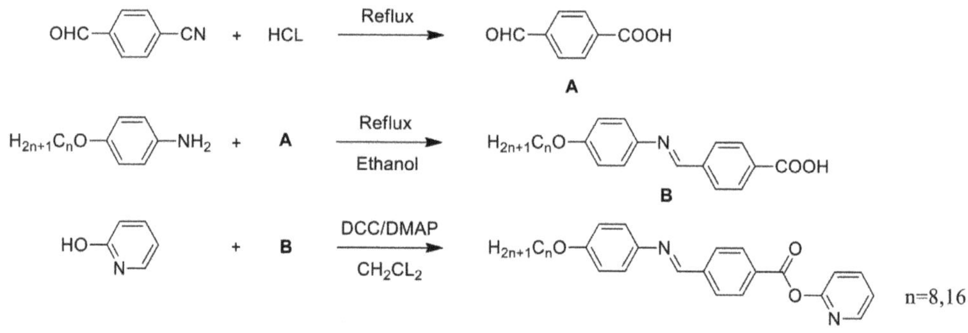

I C_8 and **I** C_{16}

Scheme 1. Synthesis of pyridin-2-yl 4-[4-(alkoxy)phenyliminomethyl]benzoate **I** $C_{(8, 16)}$.·

2.2.1. Synthesis of 4-Formylbenzoic Acid

A mixture of 4-formylbenzonitrile (1.0 g, 7.6 mmol) and hydrochloric acid (5 mL) was boiled under reflux for 4 h. The mixture was cooled at room temperature then filtered and washed with cold distilled water. The obtained solid was recrystallized from hot water (see Scheme 1, the top line).

2.2.2. Synthesis of Schiff Base Acid

Equimolar amounts of 4-formylbenzoic acid (500 mg, 3.3 mmol) and 4-alkoxyaniline (3.3 mmol) in ethanol (15 mL) were refluxed for 3 h. The progress of the reaction was monitored using thin-layer chromatography (TLC) on silica gel 60 F_{254} E-Merck (layer thickness 0.2 mm) plates with hexane-ethyl acetate (1:1 v/v) mobile phase. The spots were visualized under a UV lamp at λ = 254 nm. The resulting mixture was cooled and filtered. The product was washed with ethanol and recrystallized from hot ethanol to give a pure compound (see Scheme 1, the middle line).

2.2.3. Synthesis of Schiff Base Ester

Equimolar amounts of 4-[4-(alkoxy)phenyliminomethyl]benzoic acid (0.3 mmol) and 2-hydroxypyridine (0.3 mmol) were dissolved in dry dichloromethane (10 mL). N,N'-dicyclohexylcarbodiimide (1.2 mmol) and a few 4-dimethylaminopyridine, which was employed as a catalyst, were added to the mixture. The mixture was left under stirring for 72 h at room temperature then filtered to remove the by-product 1,3-dicyclohexyl urea (DCU). The filtrate was left until it completely evaporated then the product was recrystallized from hot ethanol to give a pure compound. The progress of the reaction was monitored using thin-layer chromatography (TLC) on silica gel 60 F_{254} E-Merck (layer thickness 0.2 mm) plates with hexane-ethyl acetate (1:1 v/v) mobile phase. The spots were visualized under a UV lamp at λ = 254 nm (see Scheme 1, the bottom line).

Pyridin-2-yl 4-[4-(octyloxy)phenyliminomethyl]benzoate (I C_8***):***

Yield: 81.0%, FTIR ($\acute{\upsilon}$, cm^{-1}): 2947–2854 (CH_2 stretching), 1728 (C=O), 1605 (C=N), 1589 (C=C), 1466 (C-O_{Asym}), 1250 (C-O_{Sym}). ^{1}H NMR (400 MHz, $CDCl_3$): δ/ppm: 0.92 (t, 3 H, J = 8 Hz, $CH_3(CH_2)_4CH_2CH_2CH_2$), 1.31–1.38 (m, 8 H, $CH_3(CH_2)_4CH_2CH_2CH_2$), 1.46–1.49 (m, 2 H, $CH_3(CH_2)_4CH_2CH_2CH_2$), 1.79–1.84 (m, 2 H, $CH_3(CH_2)_4CH_2CH_2CH_2$), 4.00 (t, 2 H, J = 8 Hz, $CH_3(CH_2)_4CH_2CH_2CH_2$), 6.97 (d, 2 H, J = 8 Hz, Ar-H), 7.27 (d, 2 H, J = 8 Hz, Ar-H), 7.38 (d, 2 H, J = 8 Hz, Ar-H), 7.51 (dd, 1 H, J_1 = 8, J_2 = 4 Hz, Py-H), 8.01 (d, 2 H, J = 8 Hz, Ar-H), 8.48–8.54 (m, 2 H, Py-H), 8.90 (dd, 1 H, J_1 = 8, J_2 = 4 Hz, Py-H), 9.44 (s, 1 H, CH=N), (Supplementary Materials).

Pyridin-2-yl 4-[4-(hexadecyloxy)phenyliminomethyl]benzoate **(I C_{16}):**

Yield: 83.0%, **FTIR ($\acute{\upsilon}$, cm^{-1}):** 2916–2854 (CH_2 stretching), 1728 (C=O), 1605 (C=N), 1589 (C=C), 1466 (C-O_{Asym}), 1250 (C-O_{Sym}). ^{1}H NMR (400 MHz, $CDCl_3$): δ/ppm: 0.92 (t, 3 H, *J* = 8 Hz, $\underline{CH_3}(CH_2)_{12}CH_2CH_2CH_2$), 1.32–1.37 (m, 24 H, $CH_3\underline{(CH_2)_{12}}CH_2CH_2CH_2$), 1.48–1.62 (m, 2 H, $CH_3(CH_2)_{12}\underline{CH_2}CH_2CH_2$), 1.81–1.84 (m, 2 H, $CH_3(CH_2)_{12}CH_2\underline{CH_2}CH_2$), 4.00 (t, 2 H, *J* = 8 Hz, $CH_3(CH_2)_{12}CH_2CH_2\underline{CH_2}$), 6.97 (d, 2 H, *J* = 8 Hz, Ar-H), 7.28 (d, 2 H, *J* = 8 Hz, Ar-H), 7.37 (d, 2 H, *J* = 8 Hz, Ar-H), 7.52 (dd, 1 H, J_1 = 8, J_2 = 4 Hz, Py-H), 8.01 (d, 2 H, *J* = 8 Hz, Ar-H), 8.48–8.53 (m, 2 H, Py-H), 8.90 (dd, 1 H, J_1 = 8, J_2 = 4 Hz, Py-H), 9.44 (s, 1 H, CH=N) (Supplementary Materials).

The phase changes in the materials were determined via differential scanning calorimetry (DSC), DSC-60A, Shimadzu, Japan. Specimens of the size 2–3 mg were encapsulated in aluminium pans and were heated or cooled under a dry nitrogen atmosphere. Measurements were performed at 10.0 °C/min. Samples were heated from room temperature to 200 °C and cooled back to room temperature at the same heating rate, all under an inert nitrogen gas atmosphere. The phase transition temperature values were determined from the endothermic peak minima of enthalpy in the heating curves. The accuracy of temperature monitoring was better than 1.0 °C.

Transition temperatures for the prepared compounds were checked and phases were identified by polarized optical microscope (POM, Wild, Germany) attached with Mettler FP82HT hot stage.

3. Results and Discussion

3.1. Mesomorphic Behaviour

The investigated compounds were characterized for their mesomorphic behavior by differential scanning calorimetry (DSC) and polarized optical microscopy (POM). The types of mesophases were identified by POM. The phase transition temperatures (°C), enthalpy of transition ΔH (kJ/mol), and normalized entropy of transition ΔS/R were measured by DSC. All the results are tabulated in Table 1 and the mesophase ranges are graphically represented in Figure 2. An indicative example demonstrating the DSC curve of the heating/cooling cycles for **I C_8** is shown in Figure 3. The presence of sharp peaks in the DSC thermogram indicates the phase changing between the crystal and the liquid crystalline. Additionally, the small peaks represent the transition from the liquid crystalline phase to the isotropic liquid. The type of mesophase texture for **I C_{16}** is represented in Figure 4.

Table 1 showed the phase transition temperatures (°C) at 10.0 °C/min, enthalpy of transition ΔH (kJ/mol), the entropy of transition ΔS (J/mol.K), and the normalized entropy of transition of compounds **I C_8**, **I C_{16}**, **II C_6**, **III C_6**, **IV C_8**, and **V C_6**. It was obvious that all the compounds have a wide range of mesophase stability, thus, they are mesogenic in nature. **IV C_8** is the only dimorphic compound that exhibited in smectic A and nematic mesophases with mesomorphic ranges of 7.1°C and 38.2 °C, respectively. The smectic A-ranges decreased in order **III C_6** (44.2 °C) > **II C_6** (20 °C) > **IV C_8** (7.1 °C). Additionally, the nematic ranges increased in order **I C_8** (14.17 °C) < **I C_{16}** (14.96 °C) < **V C_6** (27.2 °C) < **IV C_8** (38.2 °C). These trends were affected by the position and orientation of the mesogenic core. On the other hand, **V C_6** had the highest melting temperature (Cr-N) with 121.8 °C, and **II C_6** (Cr-SmA) had the lowest one with 67.3 °C. The development of SmA mesophase in **IV C_8** and not in **V C_6** was probably due to changes in the orientation of the mesogenic (-CH=N-) as it increased the dipole moment for **IV C_8** over **V C_6**. The results of the dipole moment will be discussed later as one of the important factors affecting mesomorphic behavior. On the same approach, the existence of SmA mesophase in **II C_6** and not in **I C_8** nor **I C_{16}** could be explained in terms of the dipole moment where the higher dipole moment could enhance the smectic mesophase [27,28].

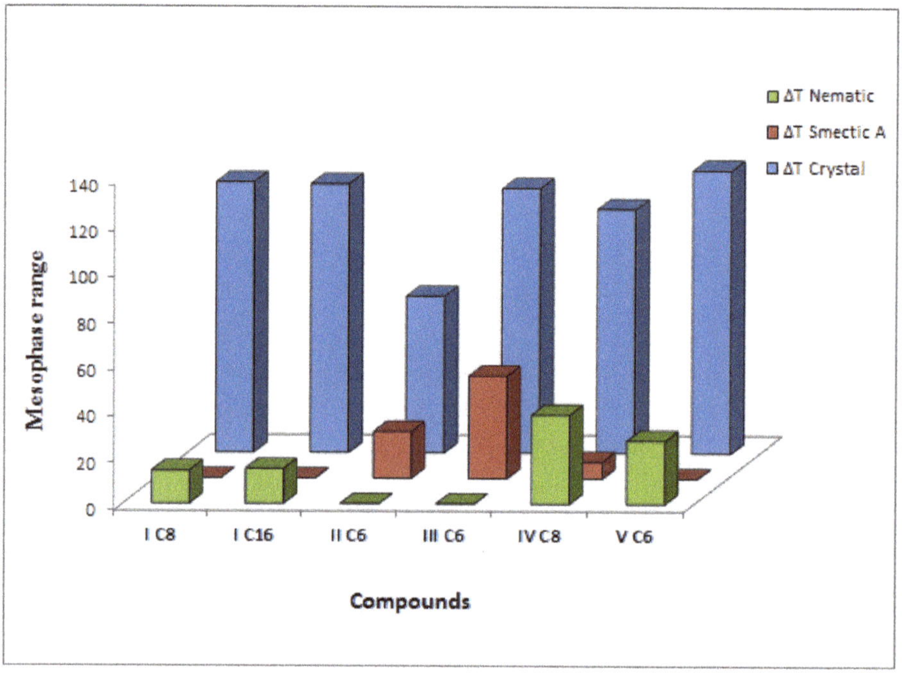

Figure 2. Mesophase ranges of compounds **I** C_8, **I** C_{16}, **II** C_6, **III** C_6, **IV** C_8, and **V** C_6.

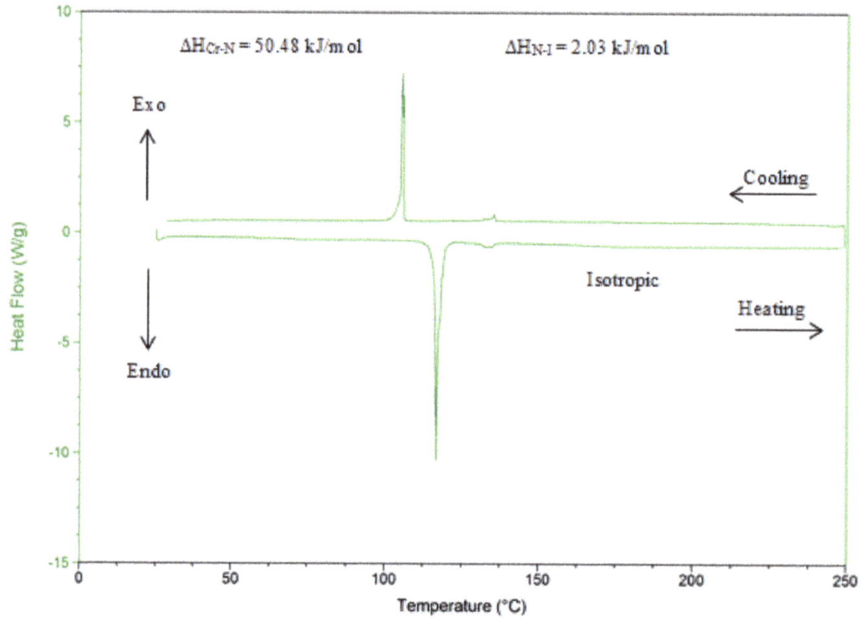

Figure 3. Differential scanning calorimetry (DSC) thermogram of compounds **I** C_8 during heating and cooling cycles, at 10.0 °C/min.

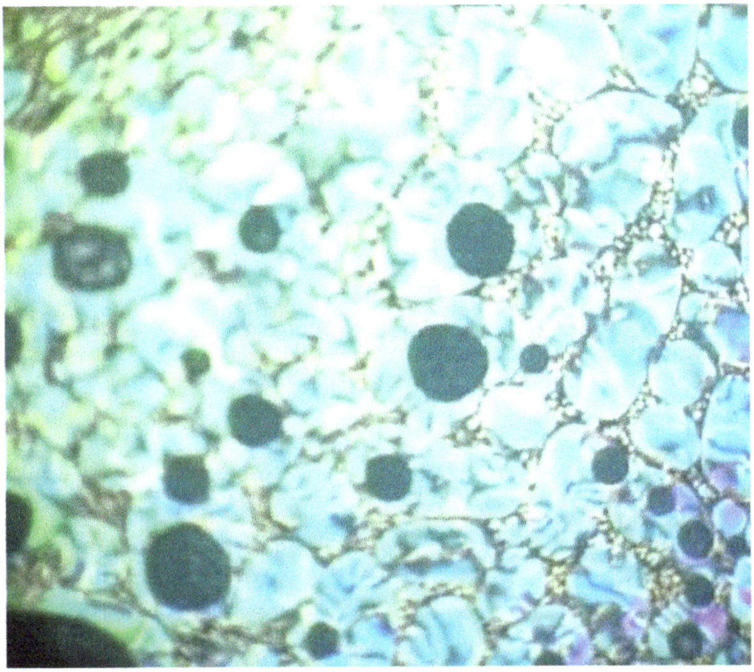

Figure 4. Nematic texture under polarized optical microscopy (POM) upon heating for **I C_{16}** at 122 °C, at 10.0 °C/min.

3.2. *DFT Calculations*

3.2.1. The Geometrical Structure

The optimum geometrical structure of all compounds was calculated using Gaussian 09 software [29] and performed using the DFT/B3LYP method using a 6–311G(d,p) basis set. The geometry for each compound was optimized to find the geometrical structure for the minimum energy of conformations regarding all geometrical parameters, then, the optimized structures were used in the estimation of the frequency. The optimized molecular geometrical structures are illustrated in Figure 5. The twist angles of compounds are tabulated in Table 2 and illustrated in Figure 6, where the red, blue and green planes are passing through the rings A, B and C, respectively. It was observed from the results of the DFT calculations of all compounds that the three rings A, B, and C were non-co-planer. The deviation from planarity could be related to the position of the mesogenic cores (-CH=N- and COO) of the liquid crystalline compounds. It is observed from Table 2 that twist angles between A and C rings in **I C_8** (6.63°) and **I C_{16}** (9.04°) compounds were much lower than their analogous phenyl compound **II C_6** (79.97°). One could conclude from the previous observation that the presence of nitrogen atoms enhanced the planarity which increased the mesophase temperature.

Table 1. Phase transition temperatures (°C) at 10.0 °C/min., enthalpy of transition ΔH (kJ/mol), the entropy of transition ΔS (J/mol.K), and normalized entropy of transition of compounds **I C_8, I C_{16}, II C_6, III C_6, IV C_8,** and **V C_6**, all values have been calculated for the heating transition.

Compounds	$T_{Cr\text{-}SmA}$	$T_{Cr\text{-}N}$	$T_{SmA\text{-}N}$	$T_{SmA\text{-}I}$	$T_{N\text{-}I}$	$\Delta H_{Cr\text{-}N}$	$\Delta S_{Cr\text{-}N}$	$\Delta S_{Cr\text{-}N}/R$	$\Delta H_{Cr\text{-}SmA}$	$\Delta S_{Cr\text{-}SmA}$	$\Delta S_{Cr\text{-}SmA}/R$	$\Delta H_{N\text{-}I}$	$\Delta S_{N\text{-}I}$	$\Delta S_{N\text{-}I}/R$	$\Delta H_{SmA\text{-}I}$	$\Delta S_{SmA\text{-}I}$	$\Delta S_{SmA\text{-}I}/R$
I C_8	-	116.42	-	-	130.59	50.48	129.58	15.58	-	-	-	2.03	5.03	0.60	-	-	-
I C_{16}	-	115.65	-	-	130.61	69.48	178.70	21.49	-	-	-	4.51	11.17	1.34	-	-	-
II C_6 [23]	67.3	-	-	87.3	-	-	-	-	28.38	83.36	10.02	-	-	-	1.92	5.33	2.65
III C_6 [24]	113.9	-	-	158.1	-	-	-	-	36.39	94.01	11.30	-	-	-	0.95	2.20	0.26
IV C_8 [25]	105.1	-	112.2	-	150.4	-	-	-	-	-	-	0.3	0.7	0.08	-	-	-
V C_6 [26]	-	121.8	-	-	149.0	-	-	-	-	-	-	0.81	1.91	0.23	-	-	-

Cr-SmA = Crystal to smectic A transition; Cr-N = Crystal to nematic transition; SmA-I = Smectic A to isotropic liquid; SmA-N = Smectic A to nematic transition; N-I = Nematic to isotropic liquid transition.

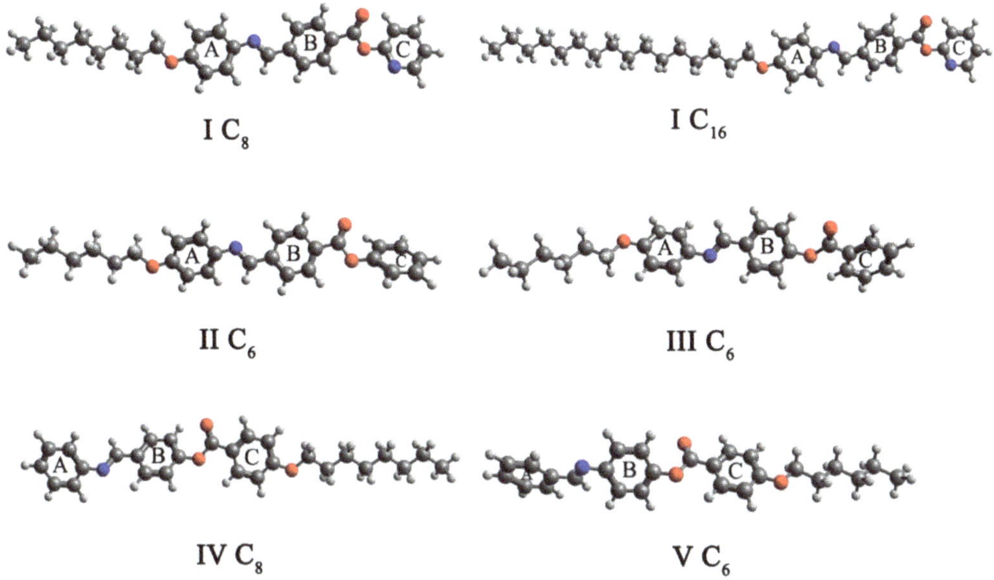

Figure 5. Calculated molecular geometrical structures of compounds **I C$_8$**, **I C$_{16}$**, **II C$_6$**, **III C$_6$**, **IV C$_8$**, and **V C$_6$** (blue—N, red—O, white—H, grey—C).

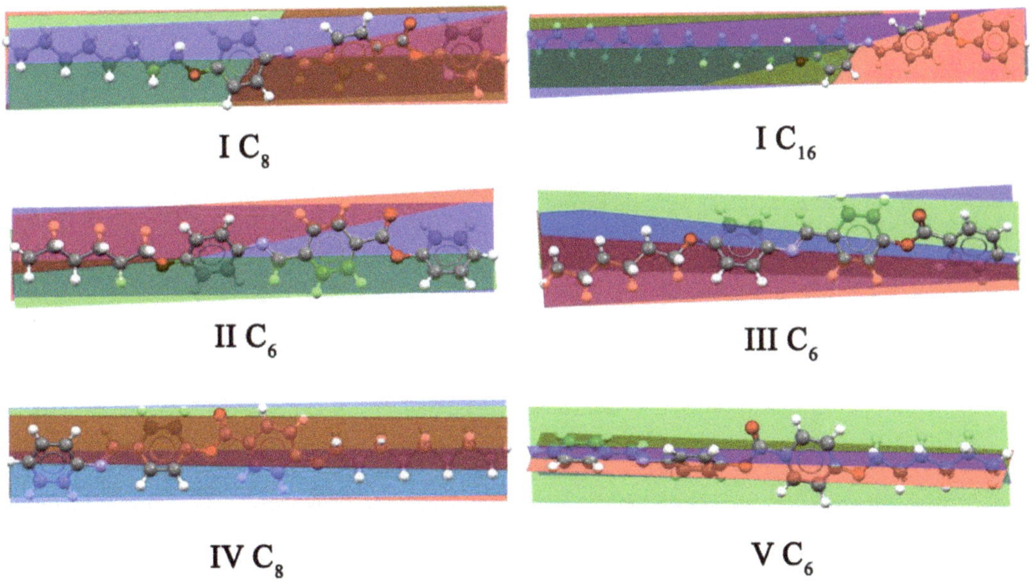

Figure 6. Estimated geometrical structure and twist angles of compounds **I C$_8$**, **I C$_{16}$**, **II C$_6$**, **III C$_6$**, **IV C$_8$**, and **V C$_6$**.

Table 2. The estimated twist angles for compounds **I C_8**, **I C_{16}**, **II C_6**, **III C_6**, **IV C_8**, and **V C_6**.

Compounds	$\theta_{A\text{-}B}$	$\theta_{A\text{-}C}$	$\theta_{B\text{-}C}$
I C_8	30.57	6.63	26.41
I C_{16}	29.32	9.04	22.96
II C_6	31.42	79.97	48.91
III C_6	34.13	75.00	40.88
IV C_8	43.78	12.21	31.59
V C_6	41.00	89.63	48.63

$\theta_{A\text{-}B}$ = Twist angle between the planes of A ring and B ring; $\theta_{A\text{-}C}$ = Twist angle between the planes of A ring and C ring; $\theta_{B\text{-}C}$ = Twist angle between the planes of B ring and C ring.

The different lengths and widths of the investigated compounds resulted in the aspect ratios and areas which are tabulated in Table 3. Aspect ratio is the ratio of width to height while the area represents the amount or extent of surface which equals to the multiply of the width by the height. These quantities demonstrated the collision diameter of the compounds. The compounds with higher aspect ratios were more likely to have stronger terminal and lateral interactions as the space-filling of the liquid crystals compounds increased. Compound C_{16}, which had the longest alkoxy chain, possessed the highest aspect ratio value. The order of increasing aspect ratios was **III C_6** < **II C_6** < **V C_6** < **I C_8** < **IV C_8** < **I C_{16}**.

Table 3. The calculated dimensions (A), area (A^2), and aspect ratio of compounds **I C_8**, **I C_{16}**, **II C_6**, **III C_6**, **IV C_8**, and **V C_6**.

Compounds	Length	Width	Height	Area	Aspect Ratio
I C_8	31.35	7.48	4.63	234.50	4.19
I C_{16}	41.60	7.74	4.83	321.98	5.38
II C_6	28.78	7.27	6.11	209.23	3.96
III C_6	28.66	7.47	5.97	214.09	3.84
IV C_8	31.52	7.50	4.72	236.40	4.20
V C_6	28.89	6.92	6.02	199.92	4.18

It was noticed that changing the orientation of the mesogenic core (-CH=N-) between **IV C_8** and **V C_6** may lead to a very small change in the aspect ratio. However, changing the position of the alkoxy terminal chain altered the aspect ratio significantly as seen in **III C_6** and **V C_6**.

It is noticed from Figure 7 that only compounds **II C_6**, **III C_6**, and **IV C_8** have a smectic A mesophase. It is worth mentioning that changing the orientation of the mesogenic core (COO) in **II C_6** and **III C_6** may alter the aspect ratio value. Compound **II C_6** had a higher aspect ratio than compound **III C_6** with the same chain length (C = 6), but the smectic A mesophase range was lower in compound **II C_6** (ΔT = 20 °C) compared to what **III C_6** had (ΔT = 44.2 °C).

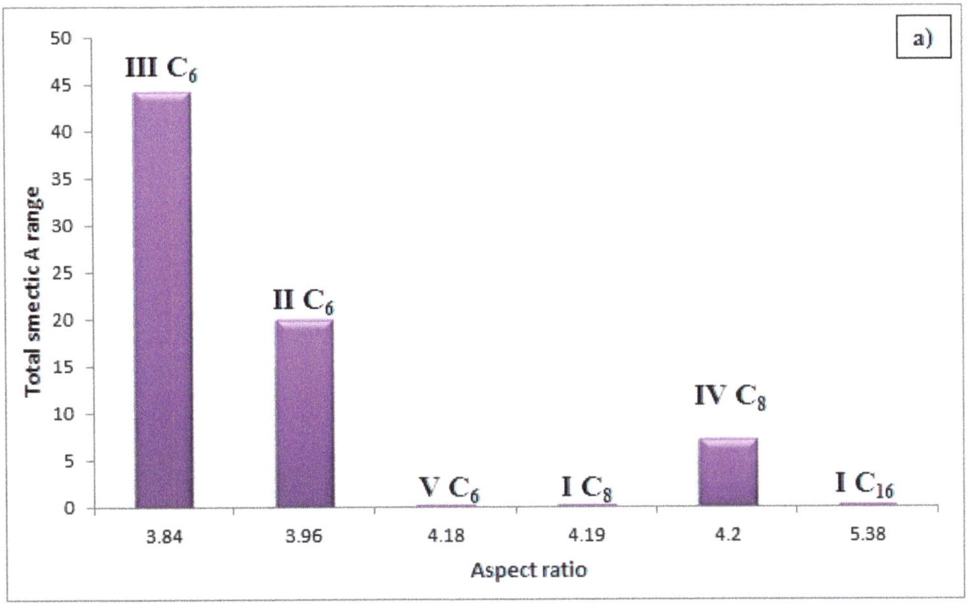

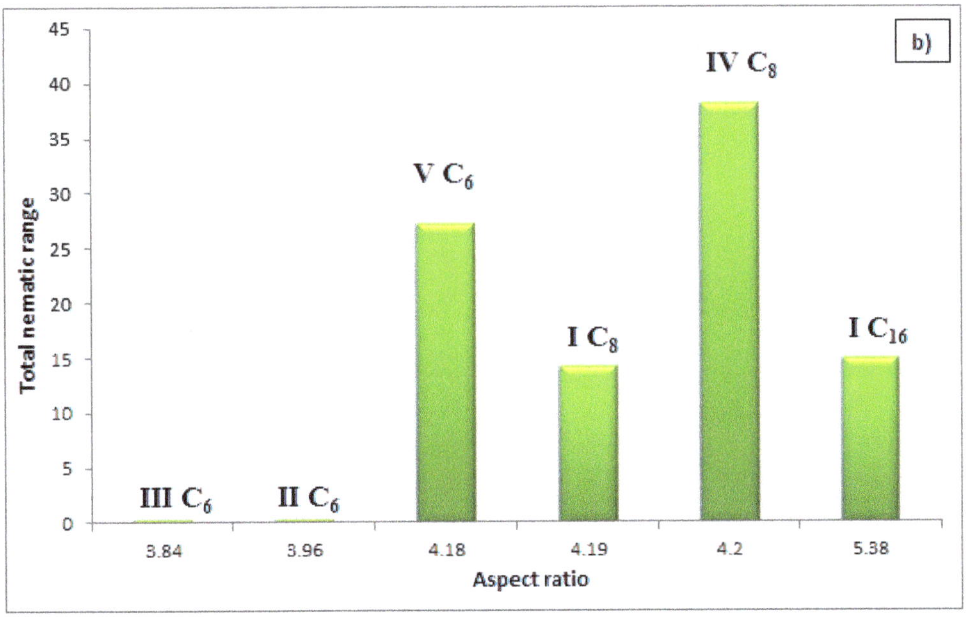

Figure 7. *Cont.*

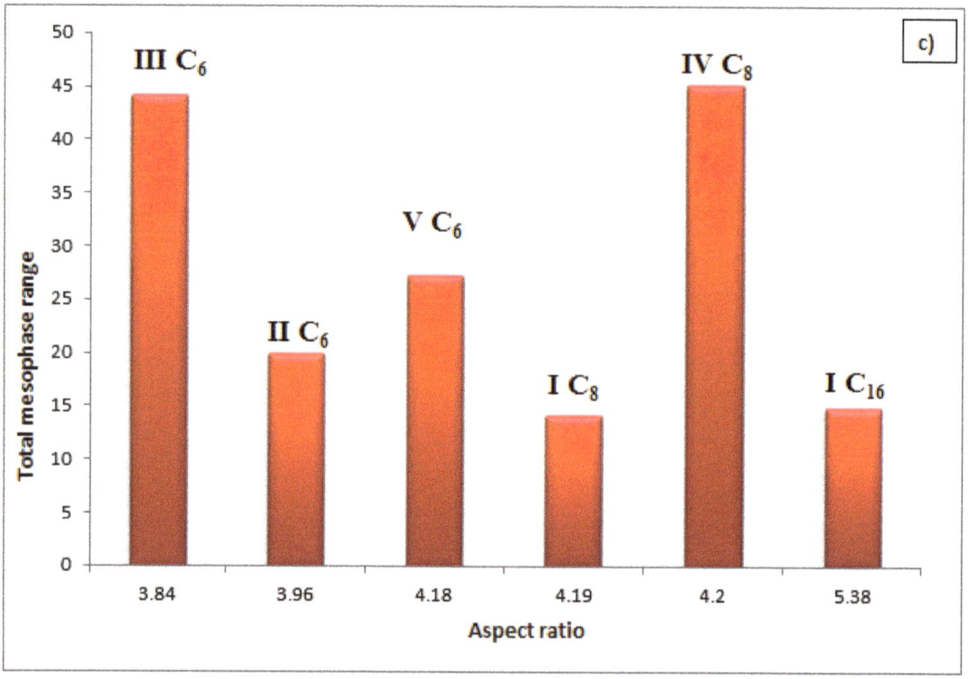

Figure 7. Dependence of aspect ratio with (**a**) Smectic range (**b**) Nematic range (**c**) Total mesophase range of compounds **I** C_8, **I** C_{16}, **II** C_6, **III** C_6, **IV** C_8, and **V** C_6.

3.2.2. Molecular Electrostatic Potentials (MEP)

Charge distribution maps were estimated according to the molecular electrostatic potential (MEP) [30] for compounds **I** C_8, **I** C_{16}, **II** C_6, **III** C_6, **IV** C_8, and **V** C_6 which were calculated by the DFT/B3LYP method using a 6–311G(d,p) basis set and provided in Figure 8. The electron density increased in the order red > orange > yellow > green > blue, therefore, the most negatively charged sites, which are illustrated in red color, were the high electronegative oxygen atoms and nitrogen atoms. The maximum was the carbonyl oxygen atom as well as the nitrogen atom in the pyridine ring (for **I** C_8 and **I** C_{16}) which will be more likely to react with the electrophilic reagent. The minimum negatively charged sites are illustrated in a blue color, where the first carbon of the terminal alkoxy chains is more probable to be attacked with nucleophiles. Expectedly, it was noticed that the length of the alkyl chain (in **I** C_8 and **I** C_{16}) had an insignificant effect on the distribution of the electron density on the remaining part of the compound. The orientation of the charge distribution could have an impact on the degree of packing of the compounds which will affect their mesophase. It was observed that the negative charge was localized in the center of compounds **III** C_6, **IV** C_8, and **V** C_6 which was expected to permit a high degree of packing. Additionally, the high degree of charge separation in **III** C_6 and **IV** C_8 compounds predicted high molecular packing which influenced the mesophase ranges.

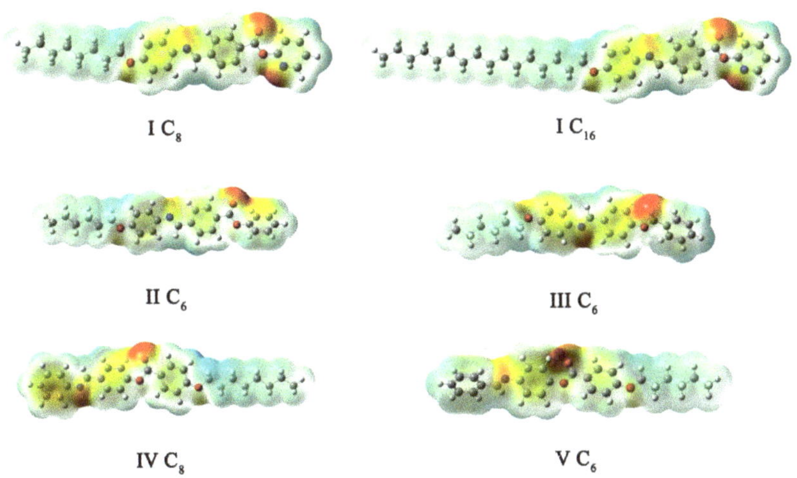

Figure 8. The estimated molecular electrostatic potentials (MEP) of compounds **I C_8**, **I C_{16}**, **II C_6**, **III C_6**, **IV C_8**, and **V C_6**.

3.2.3. Frontier Molecular Orbitals (FMOs)

Frontier molecular orbitals (FMOs) are referred to as the highest occupied molecular orbital (HOMO) and the lowest unoccupied molecular orbital (LUMO) [31]. The HOMO is an electron donor while LUMO is an electron acceptor. The frontier molecular orbitals (FMO) can predict the ability of electron transport and determine the reactivity of the molecules. The energies of HOMO and LUMO orbitals of the compounds were calculated with the same method at the same basis set and the results are tabulated in Table 4 and shown in Figure 9.

Table 4. The energies (eV), molecular orbitals energies (eV) of compounds **I C_8**, **I C_{16}**, **II C_6**, **III C_6**, **IV C_8**, and **V C_6**.

Compounds	Energy	E_{LUMO}	E_{HOMO}	$\Delta E_{(LUMO\text{-}HOMO)}$
I C_8	−37619.24	−2.30	−5.80	3.50
I C_{16}	−46179.77	−2.30	−5.80	3.50
II C_6	−35042.51	−2.28	−5.81	3.53
III C_6	−35042.52	−1.88	−5.61	3.73
IV C_8	−37182.74	−1.83	−6.03	4.20
V C_6	−35042.58	−1.81	−5.98	4.17

It is noticed that the length of the alkoxy chains in **I C_8** and **I C_{16}** had an insignificant effect on the energy gap between the FMOs. The electron densities of the regions that form the HOMO and the LUMO were mostly localized on the aromatic rings. **II C_6** of the phenyl derivative had a higher energy difference. Since the presence of the N-atom could be the factor affecting the degree of conjugation, introducing nitrogen atoms in **I C_8** and **I C_{16}** lowers the LUMO level and consequently decreased the energy gap which enhanced the electronic transition. Furthermore, it was observed that the energy gap was affected by the position of the mesogenic core (COO) as can be seen in **II C_6** and **III C_6**. The position of the carbonyl in **II C_6** may allow the maximum delocalization of the π-electrons which decreased the energy gap of the FMO. Moreover, the energy gap was higher in **IV C_8** than **III C_6** which could be attributed to the presence of higher resonance in **III C_6**, unlike **IV C_8** which may face the difficulty of resonance between the high electronegativity nitrogen atom of (-CH=N-) group and the attached aromatic ring. Additionally, there was a slight decrease in the energy gap in **V C_6** compared with **IV C_8** which may be a result of the ease of resonance between the carbon atom of the (-CH=N-) group and the aromatic ring as illustrated in Figure 10.

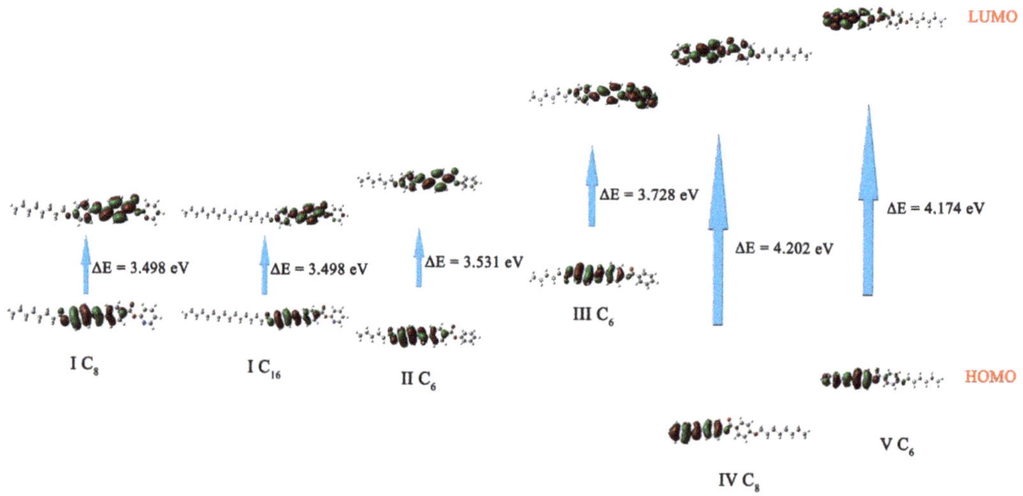

Figure 9. The estimated frontier molecular orbitals of compounds **I C_8**, **I C_{16}**, **II C_6**, **III C_6**, **IV C_8**, and **V C_6**.

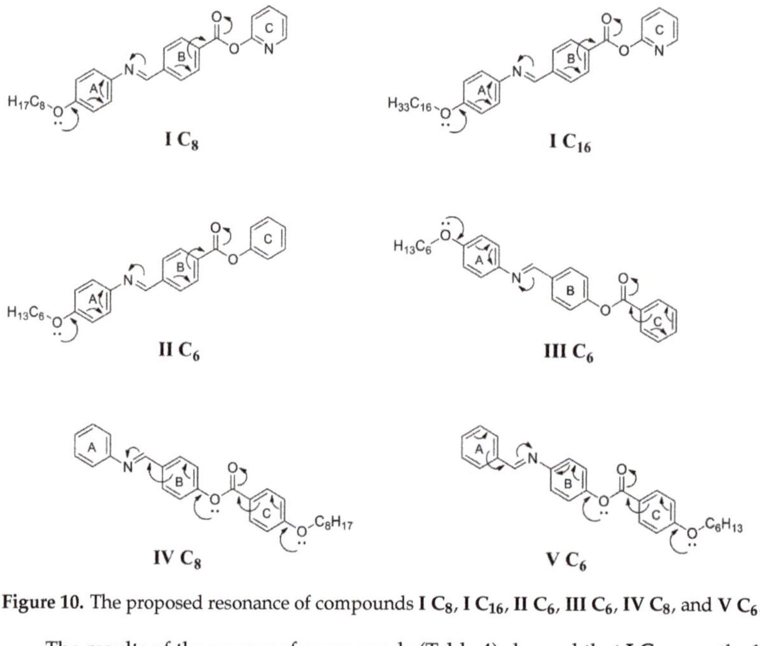

Figure 10. The proposed resonance of compounds **I C_8**, **I C_{16}**, **II C_6**, **III C_6**, **IV C_8**, and **V C_6**.

The results of the energy of compounds (Table 4) showed that **I C_{16}** was the highest stable compound. The stability of the compounds under study increased in the order **II C_6 < III C_6 < V C_6 < IV C_8 < I C_8 < I C_{16}**.

3.2.4. Dipole Moment and Polarizability

The dipole moments and polarizability were calculated with the same method at the same basis set and tabulated in Table 5. The dipole moment was greatly affected by the orientation and position of the mesogenic core as well as the distance between atoms. It was noticed that **I C_{16}** had a slightly higher dipole moment than **I C_8** which could be attributed to the longer alkyl chains of the terminal alkoxy chains. The dipole moment was increased when the same point charges were separated along a larger distance. Furthermore, it was observed that **II C_6** had a higher dipole moment than the two earlier mentioned compounds which could be related to the absence of the high electronegativity nitrogen atom of the pyridine ring thus a higher difference in electronegativity along the compound. Meanwhile, the dipole moment was dramatically decreased in **III C_6** due to the changing position of the carbonyl oxygen which could be related to the occurrence of resonance between the aromatic ring and the carbonyl oxygen that lowers the separation of charges along the compound. **IV C_8** had the highest dipole moment which could be related to the unidirectional electron delocalization by resonance along the molecular backbone. **V C_6** had a lower dipole moment due to opposing directions of resonance between central p-phenylene [ring B] with oxygen and nitrogen atoms, resulting in a cancelling effect [32,33].

Table 5. Polarizability α (a_0^3) and dipole moments (debye) of compounds **I C_8**, **I C_{16}**, **II C_6**, **III C_6**, **IV C_8**, and **V C_6**.

Compounds		I C_8	I C_{16}	II C_6	III C_6	IV C_8	V C_6
Polarizability		391.56	491.04	368.62	361.41	380.85	352.60
Dipole moment	(x)	−1.89	1.98	−2.87	0.07	4.52	−2.36
	(y)	−0.14	−0.12	−1.86	1.14	0.29	−1.40
	(z)	−1.31	1.23	0.51	0.61	−0.60	0.59
	(Total)	2.30	2.33	3.45	1.30	4.57	2.81

It is well known that the increment in the dipole moment highly affected the mesophase range [34,35]. It could be explained in terms of the degree of packing of the molecules in the liquid crystalline phases which affect the mesophase stability.

It is obvious from Figure 11 that the presence of nitrogen atoms decreased the dipole moment and mesophase range of the compounds **I C_8** and **I C_{16}**. However, the longer chain length (C = 16) increased the mesophase range without affecting its dipole moment. Its phenyl analog (**II C_6**) showed a higher dipole moment with a longer mesophase range (ΔT = 20 °C). Moreover, the absence of nitrogen atoms permited the formation of smectic mesophase that could be explained in terms of the highly ordered liquid crystalline phase that neededa greater degree of packing. As previously discussed, the orientation of the mesogenic core (-CH=N-) highly impacted the dipole moment of the compound and it also affected the degree of packing to enhance smectic mesophase for the highly polar compound (**IV C_8**). Moreover, the higher dipole moment (**IV C_8**) influenced the mesophase range to be (ΔT = 45.3 °C) with respect to (ΔT = 27.2 °C) of its analogous compound (**V C_6**). The presence of the terminal alkoxy group on the aniline ring of the compound (**III C_6**) highly decreased the dipole moment. However, the compound showed a high mesophase range (ΔT = 44.2 °C). This result could be attributed to other structural factors that could share with the small dipole moment to show such a high mesophase range.

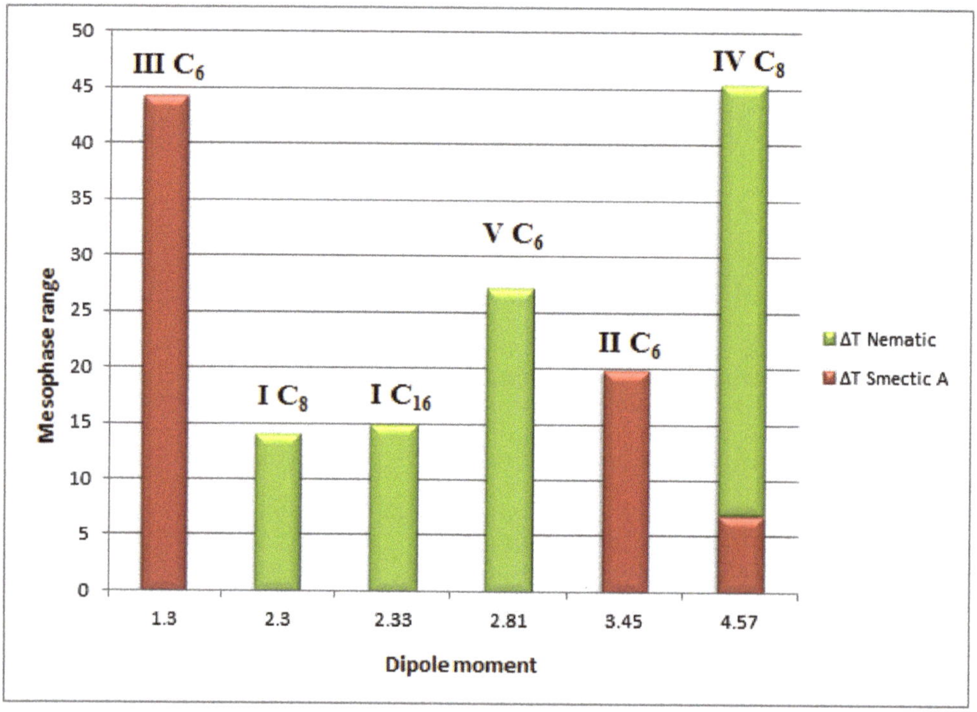

Figure 11. Graph of total mesophase range (nematic and smectic A mesophase) versus dipole moment of compounds **I C_8**, **I C_{16}**, **II C_6**, **III C_6**, **IV C_8**, and **V C_6**.

It is clear from Figure 12 that the polarizability of compound **III C_6** was more than that of compound **V C_6** with the same chain length (C=6), and such higher polarizability enhanced the mesophase range with the formation of smectic mesophase for compound **III C_6**. On the other hand, the orientation of the mesogenic core highly affected the polarizability to show higher polarizability with a lower mesophase range. The orientation of the carboxylate group (COO) obviously had a high impact on both the polarizability and the mesomorphic behavior. Compound **II C_6** had lower polarizability than compound **IV C_8** with different mesomorphic behavior. Additionally, the presence of nitrogen atoms in the terminal ring of compound **I C_8** increased the polarizability compared to its analogous compound **II C_6** but it also decreased the mesophase range and the type of texture was changed from smectic A (**II C_6**) to nematic (**I C_8**).

3.2.5. Aromaticity, LOL-π and π-π Stacking

Multiwfn software in its 3.8 version[36] was used to study aromaticity, π-π stacking of all aromatic rings of the concerned compounds according to the optimized geometries obtained from Gaussian 09 software. Aromaticity was analyzed by the means of normalized multicenter bond order (MCBO index) which indicates the electron delocalization capability over a ring where a larger MCBO value refers to stronger aromaticity [37]. The extended π-π stacking was evaluated using the localized orbital locator integrated π over the plane (LOLIPOP index) where the higher LOLIPOP index value of a ring indicated a weaker π-stacking capability. The values of the MCBO index and the LOLIPOP index for all rings are tabulated in Table 6. Furthermore, the isosurface of the LOL purely contributed by π-orbitals (LOL-π) [38] was estimated using Multiwfn software then visualized using VMD software [39] to render high-quality figures as illustrated in Figure 13.

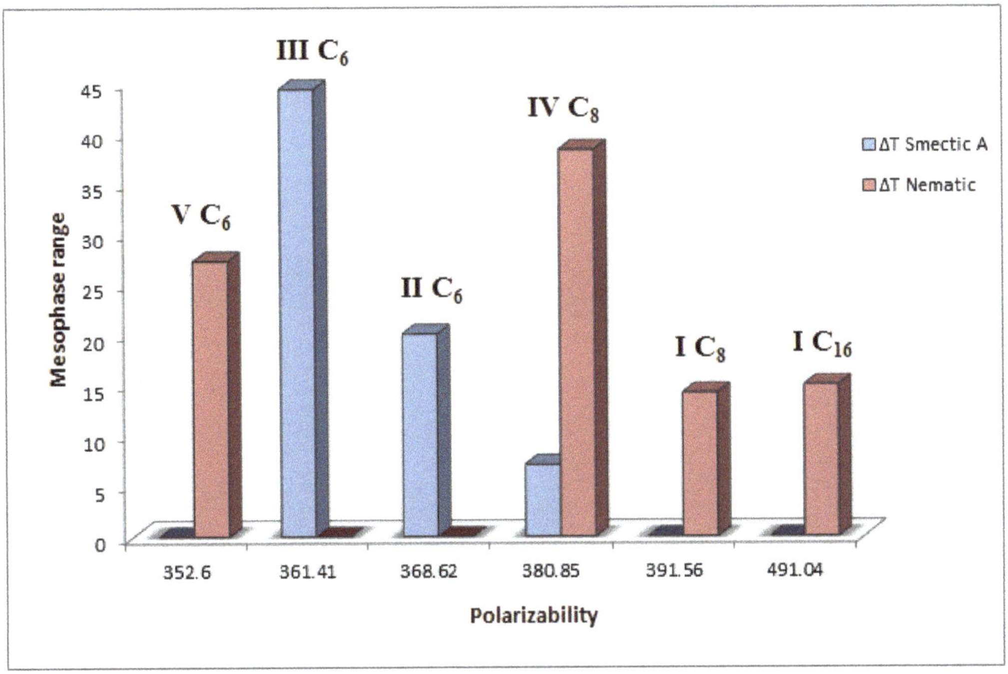

Figure 12. Graph of smectic A and nematic mesophase range versus polarizability of compounds **I** C_8, **I** C_{16}, **II** C_6, **III** C_6, **IV** C_8, and **V** C_6.

Table 6. The normalized multicenter bond order (MCBO) index and localized orbital locator integrated π over the plane (LOLIPOP) of compounds **I** C_8, **I** C_{16}, **II** C_6, **III** C_6, **IV** C_8, and **V** C_6.

Compounds	MCBO Index			LOLIPOP Index		
	A	B	C	A	B	C
I C_8	0.624	0.629	0.643	6.751	6.691	5.041
I C_{16}	0.624	0.629	0.643	6.740	6.703	5.055
II C_6	0.624	0.630	0.647	6.741	6.685	7.253
III C_6	0.626	0.632	0.643	6.814	6.777	7.135
IV C_8	0.640	0.632	0.626	7.275	6.954	7.033
V C_6	0.640	0.634	0.626	7.171	7.021	7.031

It was noticed from the normalized MCBO aromaticity index that A rings bounded to alkoxy chain in **I** C_8, **I** C_{16} and **II** C_6 have the same aromaticity (0.624). **III** C_6 had a slightly higher value (0.626) than the previously mentioned compounds. One could deduce that A rings in these compounds had almost the same degree of cyclic delocalization. Meanwhile, A rings of **IV** C_8 and **V** C_6 show an identical greater value (0.640) than other compounds which can be attributed to higher electron delocalization in the free benzene ring. Central B rings of **I** C_8, **I** C_{16} and **II** C_6 exhibited nearly similar aromatic character which was slightly lower than the aromaticity of **III** C_6 and **IV** C_8. The B ring of **V** C_6 had the highest aromaticity index which may be because of a lack of extending π-conjugation outside the ring. C rings of **I** C_8 and **I** C_{16} had identical aromaticity (0.643) which was lower than **II** C_6 (0.647) as the pyridine ring had a lower aromaticity character than benzene. It was observed that the C ring of **III** C_6 had a greater value than what **IV** C_8 and **V** C_6 had as the earlier one was free benzene.

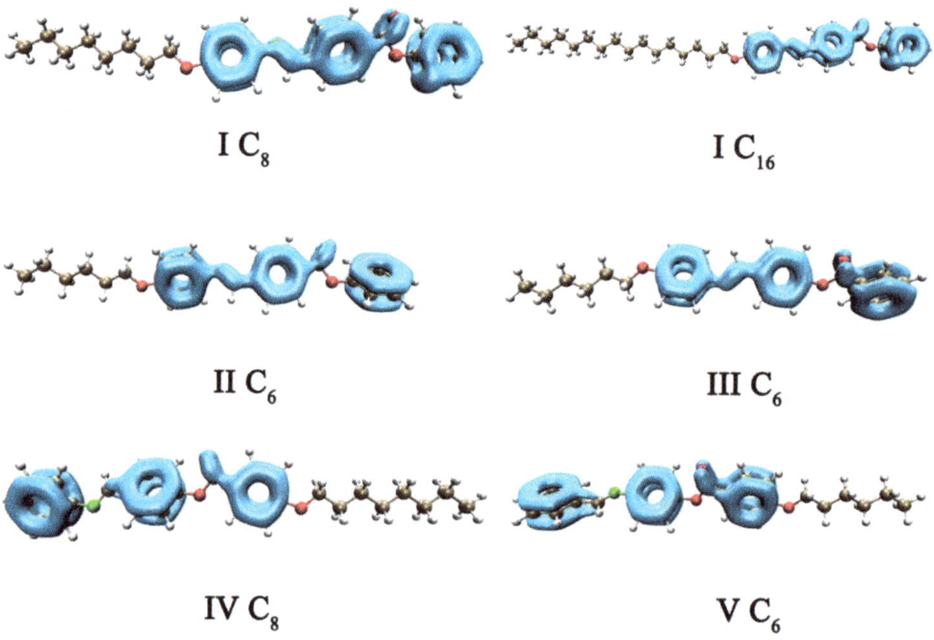

Figure 13. Isosurface maps of LOL- π (Isovalue = 0.3) of compounds **I C$_8$**, **I C$_{16}$**, **II C$_6$**, **III C$_6$**, **IV C$_8$**, and **V C$_6$**.

It was revealed from the LOLIPOP index that inner benzene rings which are bounded to the alkoxy chain had stronger π-π stacking ability than terminal rings as seen in the C rings of **IV C$_8$** and **V C$_6$**, and also observed in the A rings of **II C$_6$** and **III C$_6$**. On the other hand, the C rings of **I C$_8$** and **I C$_{16}$** had the strongest π-π stacking ability which could be referred to as the presence of N atom.

It was found from LOL-π isosurfaces that benzene rings exert extending π-conjugation when bounded with the carbonyl group (C=O). It is also worth mentioning that the A ring in **IV C$_8$** did not exhibit extending π-conjugation due to the electronegativity of the N atom thus, there was no resonance between the imine group and the benzene ring. However, the π-conjugation extended from the A ring to the carbon of the imine group in **V C$_6$** which was probably due to the existence of resonance between the imine group and the benzene ring as the direction of electron-withdrawing was in line with the direction of resonance.

4. Conclusions

Herein liquid crystalline pyridyl analog has been prepared and fully characterized. The mesomorphic results of the prepared compound showed a nematic texture with a medium range. The investigated compounds were compared with respect to their phenyl analog to reveal that the phenyl analog has a smectic texture which could be attributed to its high dipole moment and low polarizability compared to what the pyridyl analog has. The energy gap between the FMO of pyridyl analog is lower than the energy gap of its phenyl analog which enhances the electronic transition. Moreover, the orientation of the mesogen showed a significant effect on the mesomorphic behavior where the CH=N and COO group's orientation changed the texture of the investigated compounds from nematic to smectic or calamitic. Besides, it enhanced the mesomorphic range to 44.2 °C for compound **III C$_6$** instead of 20 °C for compound **II C$_6$**. Additionally, it developed a smectic mesophase in **IV C$_8$** which did not exist in **V C$_6$**. Twist angles in **I C$_8$** and **I C$_{16}$** compounds are lower than their analogous phenyl compound **II C$_6$** which could indicate that the presence of nitrogen atom enhances the planarity which increases the mesophase temperature. In the aspect ratio approach, changing the position of the alkoxy terminal chain alters the aspect ratio significantly as seen in **III** $\mathbf{C_6}$ and **V** $\mathbf{C_6}$, Additionally, changing the orientation of the mesogenic core (COO) in **II** $\mathbf{C_6}$ and **III** $\mathbf{C_6}$ may alter the aspect ratio value. The molecular electrostatic potential implied that the high degree of charge separation predicts high molecular packing which influences the mesophase ranges. It was found using LOLIPOP index values that the terminal ring of the pyridyl analog has a strong π-π stacking ability compared with its phenyl analog. Correspondingly in the phenyl analog case, the inner benzene rings which are bound to the alkoxy chain have a stronger π-π stacking ability than the terminal rings. Furthermore, the MCBO aromaticity index suggests that the free benzene rings which are not bound to the alkoxy chain have higher aromaticity than the inner benzene rings.

Supplementary Materials: The following supporting information can be downloaded at: https://www.mdpi.com/article/10.3390/molecules27072304/s1, Figure S1: 1HNMR of Pyridin-2-yl 4-[4-(hexyloxy)phenyliminomethyl]benzoate (I C6); Figure S2: 1HNMR of Pyridin-2-yl 4-[4-(hexadecyloxy)phenyliminomethyl]benzoate (I C16).

Author Contributions: Conceptualization, M.H. and A.A.H.; Formal analysis, S.N., M.H., A.A.S. and M.A.Z.; Funding acquisition, A.-H.E. and M.J.; Investigation, S.N., A.A.H., A.-H.E. and M.A.Z.; Methodology, S.N. and M.A.Z.; Project administration, A.-H.E. and M.J.; Resources, A.-H.E. and M.J.; Software, S.N. and A.A.H.; Supervision, M.H. and O.F.; Writing—original draft, S.N., M.H. and A.A.H.; Writing—review & editing, S.N., M.H., A.A.H. and M.A.Z. All authors have read and agreed to the published version of the manuscript.

Funding: Smart-Health Initiative (SHI) and Red Sea Research Center (RSRC), Division of Biological and Environmental Sciences and Engineering (BESE), King Abdullah University of Science and Technology (KAUST), Thuwal, Jeddah, Saudi Arabia—authors are thankful for funding this project either chemicals or analysis.

Institutional Review Board Statement: Not applicable.

Informed Consent Statement: Not applicable.

Data Availability Statement: Not applicable.

Conflicts of Interest: The authors declare no conflict of interest.

Sample Availability: Samples of the compounds are available from the authors.

References

1. Dierking, I.; Al-Zangana, S. Lyotropic liquid crystal phases from anisotropic nanomaterials. *Nanomaterials* **2017**, *7*, 305. [CrossRef]
2. Lagerwall, J.P.; Scalia, G. A new era for liquid crystal research: Applications of liquid crystals in soft matter nano-, bio- and microtechnology. *Curr. Appl. Phys.* **2012**, *12*, 1387–1412. [CrossRef]
3. An, J.-G.; Hina, S.; Yang, Y.; Xue, M.; Liu, Y. Characterization of liquid crystals: A literature review. *Rev. Adv. Mater. Sci.* **2016**, *44*, 398–406.
4. Gray, G.W.; Vill, V.; Spiess, H.W.; Demus, D.; Goodby, J.W. *Physical Properties of Liquid Crystals*; John Wiley & Sons: Hoboken, NJ, USA, 2009.
5. Kumar, S. *Experimental Study of Physical Properties and Phase Transitions*; Cambridge University Press: Cambridge, UK, 2001.
6. Zhang, M.; Jang, C.-H. Liquid crystal based optical sensor for imaging trypsin activity at interfaces between aqueous phases and thermotropic liquid crystals. *Bull. Korean Chem. Soc.* **2013**, *34*, 2973–2977. [CrossRef]
7. Sergeyev, S.; Pisula, W.; Geerts, Y.H. Discotic liquid crystals: A new generation of organic semiconductors. *Chem. Soc. Rev.* **2007**, *36*, 1902–1929. [CrossRef] [PubMed]
8. O'Neill, M.; Kelly, S.M. Ordered materials for organic electronics and photonics. *Adv. Mater.* **2011**, *23*, 566–584. [CrossRef] [PubMed]
9. Yu, Y.; Nakano, M.; Ikeda, T. Directed bending of a polymer film by light. *Nature* **2003**, *425*, 145. [CrossRef]
10. Thomsen, D.L.; Keller, P.; Naciri, J.; Pink, R.; Jeon, H.; Shenoy, D.; Ratna, B.R. Liquid crystal elastomers with mechanical properties of a muscle. *Macromolecules* **2001**, *34*, 5868–5875. [CrossRef]
11. Rananavare, S.B.; Pisipati, V. An Overview of Liquid Crystals Based on Schiff Base Compounds. In *Liquid Crystalline Organic Compounds and Polymers as Materials of the XXI Century: From Synthesis to Applications*; Iwan, A., Schab-Balcerzak, E., Eds.; Transworld Research Network: Trivandrum, India, 2011.

12. Kelker, H.; Scheurle, B. A liquid-crystalline (nematic) phase with a particularly low solidification point. *Angew. Chem. Int. Ed. Engl.* **1969**, *8*, 884–885. [CrossRef]
13. Hallsby, A.; Nilsson, M.; Otterholm, B. Synthesis of Schiff Bases Forming the First Room Temperature Ferroelectric Liquid Crystal—The Mora Series. *Mol. Cryst. Liq. Cryst.* **1982**, *82*, 61–68. [CrossRef]
14. Gomha, S.M.; Ahmed, H.A.; Shaban, M.; Abolibda, T.Z.; Khushaim, M.S.; Alharbi, K.A. Synthesis, optical characterizations and solar energy applications of new Schiff base materials. *Materials* **2021**, *14*, 3718. [CrossRef]
15. Wu, J.-N.; Chen, L.; Fu, T.; Zhao, H.-B.; Guo, D.-M.; Wang, X.-L.; Wang, Y.-Z. New application for aromatic Schiff base: High efficient flame-retardant and anti-dripping action for polyesters. *Chem. Eng. J.* **2018**, *336*, 622–632. [CrossRef]
16. Yeap, G.-Y.; Ha, S.-T.; Boey, P.-L.; Mahmood, W.A.K.; Ito, M.M.; Youhei, Y. Synthesis and characterization of some new mesogenic schief base esters N-[4-(4-n-hexadecanoyloxybenzoyloxy)-benzylidene]-4-substituted anilines. *Mol. Cryst. Liq. Cryst.* **2006**, *452*, 73–90. [CrossRef]
17. Ahmed, H.; Hagar, M.; El-Sayed, T.; Alnoman, R.B. Schiff base/ester liquid crystals with different lateral substituents: Mesophase behaviour and DFT calculations. *Liq. Cryst.* **2019**, *46*, 1–11. [CrossRef]
18. Ahmed, H.; Hagar, M.; Saad, G. Impact of the proportionation of dialkoxy chain length on the mesophase behaviour of Schiff base/ester liquid crystals; experimental and theoretical study. *Liq. Cryst.* **2019**, *46*, 1611–1620. [CrossRef]
19. Alnoman, R.; Ahmed, H.A.; Hagar, M. Synthesis, optical, and geometrical approaches of new natural fatty acids' esters/Schiff base liquid crystals. *Molecules* **2019**, *24*, 4293. [CrossRef]
20. Nafee, S.S.; Hagar, M.; Ahmed, H.A.; Alhaddad, O.; El-Shishtawy, R.M.; Raffah, B.M. New two rings Schiff base liquid crystals; ball mill synthesis, mesomorphic, Hammett and DFT studies. *J. Mol. Liq.* **2020**, *299*, 112161. [CrossRef]
21. Ahmed, N.H.; Saad, G.R.; Ahmed, H.A.; Hagar, M. New wide-stability four-ring azo/ester/Schiff base liquid crystals: Synthesis, mesomorphic, photophysical, and DFT approaches. *RSC Adv.* **2020**, *10*, 9643–9656. [CrossRef]
22. Hagar, M.; Ahmed, H.; Saad, G. New calamitic thermotropic liquid crystals of 2-hydroxypyridine ester mesogenic core: Mesophase behaviour and DFT calculations. *Liq. Cryst.* **2020**, *47*, 114–124. [CrossRef]
23. Nafee, S.S.; Hagar, M.; Ahmed, H.A.; El-Shishtawy, R.M.; Raffah, B.M. The synthesis of new thermal stable schiff base/ester liquid crystals: A computational, mesomorphic, and optical study. *Molecules* **2019**, *24*, 3032. [CrossRef]
24. Hagar, M.; Ahmed, H.; Aouad, M. Mesomorphic and DFT diversity of Schiff base derivatives bearing protruded methoxy groups. *Liq. Cryst.* **2020**, *47*, 2222–2233. [CrossRef]
25. Hagar, M.; Ahmed, H.; Saad, G. Mesophase stability of new Schiff base ester liquid crystals with different polar substituents. *Liq. Cryst.* **2018**, *45*, 1324–1332. [CrossRef]
26. Hagar, M.; Ahmed, H.; Saad, G. Synthesis and mesophase behaviour of Schiff base/ester 4-(arylideneamino) phenyl-4"-alkoxy benzoates and their binary mixtures. *J. Mol. Liq.* **2019**, *273*, 266–273. [CrossRef]
27. Yamamura, Y.; Adachi, T.; Miyazawa, T.; Horiuchi, K.; Sumita, M.; Massalska-Arodź, M.; Urban, S.; Saito, K. Calorimetric and spectroscopic evidence of chain-melting in smectic e and smectic a phases of 4-alkyl-4′-isothiocyanatobiphenyl (n TCB). *J. Phys. Chem. B* **2012**, *116*, 9255–9260. [CrossRef] [PubMed]
28. Wunderlich, B. A classification of molecules, phases, and transitions as recognized by thermal analysis. *Thermochim. Acta* **1999**, *340*, 37–52. [CrossRef]
29. Frisch, M.; Trucks, G.; Schlegel, H.B.; Scuseria, G.E.; Robb, M.A.; Cheeseman, J.R.; Scalmani, G.; Barone, V.; Mennucci, B.; Petersson, G. *Gaussian 09, Revision D. 01*; Gaussian. Inc.: Wallingford, CT, USA, 2009; Volume 201.
30. Romeo, M. Density–orientation coupling for a microcontinuum approach to nematic liquid crystals subject to electric field. *Contin. Mech. Thermodyn.* **2021**, *33*, 835–849. [CrossRef]
31. Chen, R.; Wang, L.; An, Z.; Chen, X.; Chen, P. Effect of π-conjugation units on the liquid crystal and photovoltaic performance of heterocyclic pyridine-based compounds. *Liq. Cryst.* **2021**, *48*, 2178–2187. [CrossRef]
32. Li, J.; Xia, R.; Xu, H.; Yang, J.; Zhang, X.; Kougo, J.; Lei, H.; Dai, S.; Huang, H.; Zhang, G. How Far Can We Push the Rigid Oligomers/Polymers toward Ferroelectric Nematic Liquid Crystals? *J. Am. Chem. Soc.* **2021**, *143*, 17857–17861. [CrossRef]
33. Jain, V.; Kaur, S.; Mohiuddin, G.; Pal, S.K. Design, Synthesis and Characterization of Achiral Unsymmetrical Four-Ring Based Hockey-Stick Shaped Liquid Crystals: Structure-Property Relationship. *Liq. Cryst.* **2021**, 1–10. [CrossRef]
34. Goodby, J.W. Free volume, molecular grains, self-organisation, and anisotropic entropy: Machining materials. *Liq. Cryst.* **2017**, *44*, 1755–1763. [CrossRef]
35. Singh, S. *Liquid Crystals: Fundamentals*; World Scientific: Singapore, 2002.
36. Lu, T.; Chen, F. Multiwfn: A multifunctional wavefunction analyzer. *J. Comput. Chem.* **2012**, *33*, 580–592. [CrossRef]
37. Giambiagi, M.; de Giambiagi, M.S.; Mundim, K.C. Definition of a multicenter bond index. *Struct. Chem.* **1990**, *1*, 423–427. [CrossRef]
38. Lu, T.; Chen, Q. A simple method of identifying π orbitals for non-planar systems and a protocol of studying π electronic structure. *Theor. Chem. Acc.* **2020**, *139*, 25. [CrossRef]
39. Humphrey, W.; Dalke, A.; Schulten, K. VMD: Visual molecular dynamics. *J. Mol. Graph.* **1996**, *14*, 33–38. [CrossRef]

molecules

MDPI

Article

Confinement-Induced Fabrication of Liquid Crystalline Polymeric Fibers

Jae Gwang Kim [1], Jae Gyeong Lee [2,3] and Jeong Jae Wie [1,2,3,*]

1 Department of Polymer Science and Engineering, Inha University, Incheon 22212, Korea
2 Program in Environmental and Polymer Engineering, Inha University, Incheon 22212, Korea
3 Department of Organic and Nano Engineering, Hanyang University, Seoul 04763, Korea
* Correspondence: wie@inha.ac.kr or jjwie@hanyang.ac.kr; Tel.: +82-32-860-7484

Abstract: In aqueous media, liquid crystalline droplets typically form spherical shapes in order to minimize surface energy. Recently, non-spherical geometry has been reported using molecular self-assembly of surfactant-stabilized liquid crystalline oligomers, resulting in branched and randomly oriented filamentous networks. In this study, we report a polymerization of liquid crystalline polymeric fibers within a micro-mold. When liquid crystal oligomers are polymerized in freely suspended aqueous media, curvilinear and randomly networked filaments are obtained. When reactive liquid crystalline monomers are oligomerized in a micro-channel, however, highly aligned linear fibers are polymerized. Within a top-down microfabricated mold, a bottom-up molecular assembly was successfully achieved in a controlled manner by micro-confinement, suggesting a unique opportunity for the programming architecture of materials via a hybrid approach.

Keywords: liquid crystal; liquid crystal polymers; confinement; fibers

1. Introduction

Liquid crystalline polymers have been a point of great interest in soft actuators and soft robotic applications [1]. In liquid crystal molecules, they are composed of mesogenic units and spacer units. In mesogenic units, anisotropic rod-like rigid benzene rings are able to induce intermolecular π-π interactions. As π-π interactions are directional attractive forces, the presence of only mesogenic units results in a crystal phase relative to the materials [2,3]. Conversely, the presence of only aliphatic spacer units tends to result in a liquid phase. Thus, it is plausible that materials could remain in a liquid crystalline phase with an appropriate balance between mesogenic and spacer units. Since liquid crystalline molecules have both segmental mobility and directional attractive forces, liquid crystalline materials would have the capacity for molecular self-assembly.

The ability of liquid crystalline materials in self-assembling provides programmability of molecular orientation. Due to this molecular programmability, liquid crystalline molecules have been utilized in various applications, one of the most successful commercialized products being liquid crystal displays (LCDs). In LCDs, twisted nematic (TN) geometry has often been employed where liquid crystal molecules have orthogonality between the top and bottom layers with a 90° twist through the thickness direction. Since liquid crystalline molecules are able to self-assemble between two programmed command surfaces, programming is only required in order for the surfaces to construct this TN molecular geometry; it is not necessary to program individual molecules. In addition, to follow the molecular direction at the interface of command surfaces, liquid crystal molecules gradually rotate 90° between the two layers. Despite the programmability of the molecular alignment, liquid crystal molecules have liquid-like mobility such that TN geometry is not permanent. When terminal groups of liquid crystalline molecules are functionalized (i.e., acrylates), however, the programmed geometry of liquid crystal molecules can be solidified via polymerization and thereby become fixed. For polymerization, photopolymerization is often adopted to maintain directional molecular geometry, as thermal energy causes random directional Brownian motions for thermal polymerization. Through self-assembly and photopolymerization of liquid crystalline monomers, liquid crystalline polymers with TN-derivatives have been utilized in various soft actuators and soft robotics [4–18].

When liquid crystalline molecules are freely dispersed in hydrophilic media (i.e., deionized water) without any command surfaces, their phase is separated from the hydrophilic media and forms spherical droplets. As liquid crystals mainly consist of hydrocarbons, they become hydrophobic and self-assemble with each other in order to minimize contact areas with hydrophilic media. To minimize surface tension, spherical micelles have been widely reported for the geometry of liquid crystals and liquid crystalline polymer droplets [19–22].

Recently, Shu Yang et al. reported the synthesis of fibrous liquid crystalline polymers in an aqueous solution with the presence of surfactants [23]. As is well known, surfactants have both hydrophilic head groups and hydrophobic tail groups. The addition of surfactants decreases the surface tension of water until the critical micelle concentration (CMC) is reached. Above the CMC, surface tensions no longer change by forming spherical micelles. Within the micelle, hydrophobic tail groups self-assemble at the core, surrounded by hydrophilic head groups, which are already in contact with water media. Similarly, hydrophobic liquid crystal molecules are solubilized within the various shape of micelles and self-assemble with each other, balancing between surface tension and internal elastic energy from liquid crystal molecules. At high temperatures, the micelle presented spherical shapes above the CMC because surface tension is larger than the bulk elastic energy of liquid crystal molecules. Upon cooling, however, the balance between surface tension and the elastic energy of the bulk becomes destabilized, facilitating the spontaneous formation of branched filamentous liquid crystalline monomer micelles.

Herein, we report a polymerization of highly aligned fibrous liquid crystalline polymers within a micro-channel. To prevent random directional fibril-like geometry via bottom-up self-assembly, we confined liquid crystalline monomers within a top-down micro-fabricated channel. This hybrid approach has the potential to program the architecture of liquid crystalline materials to obtain non-conventional, highly ordered fibril-like polymers without a high-shear spinning process.

2. Results and Discussion

Liquid crystalline emulsions were prepared by mixing liquid crystalline monomers (1,4-bis-[4-(6-acryloyloxyhexyloxy)benzoyloxy]-2-methylbenzene; RM82; Synthon Chemicals), chain extenders (n-butylamine; Sigma Aldrich), and photo-initiators (2-Benzyl-2-dimethylamino-1-(4-morpholinophenyl)-butanone-1; I-369; Ciba). The chemical structures of materials used in this study are shown in Figure 1. The liquid crystalline monomer and chain extender are first dissolved in chloroform. Then, liquid crystalline emulsions were constructed by mixing the solution in aqueous media with the presence of surfactants (sodium dodecyl sulfate; SDS; Sigma Aldrich). Due to the presence of surfactants, spherical liquid crystalline micelles are formed in the emulsion, generating a gradient of elastic energy density in the radial direction of the spherical micelles. Here, the core has higher elastic energies than that of the outer part. At this spherical micelle, interfacial tension, γ, is higher than bulk director elastic energy [23].

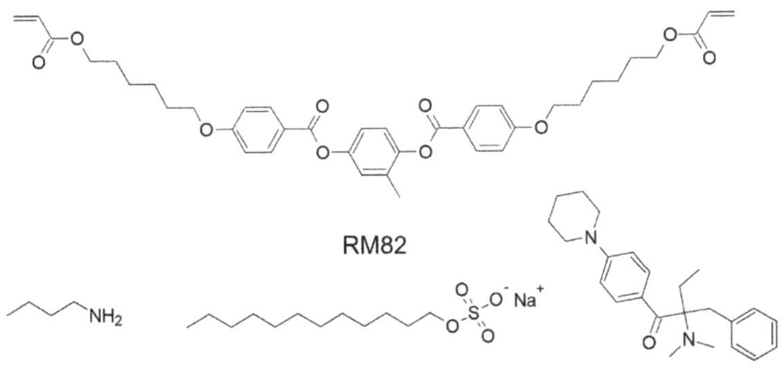

n-butylamine SDS (Sodium Dodecyl Sulfate) I-369

Figure 1. Molecular structures of chemical components were utilized in this study. 1,4-bis-[4-(6-acryloyloxyhexyloxy)benzoyloxy]-2-methylbenzene (RM82), n-butylamine, sodium dodecyl sulfate (SDS), and 2-benzyl-2-dimethylamino-1-(4-morpholinophenyl)-butanone-1 (I-369).

The RM82 liquid crystalline monomer has diacrylate functional groups and is thermally oligomerized with n-butyl amine by Michael addition reaction [24] in the micelle. With a longer total oligomerization time, the mean oligomer chain length, $<\ell>$, will increase (Figure 2a), and the elastic constant also increases following increases in molecular lengths [25]. The longer $<\ell>$ with a higher elastic constant not only increases molecular ordering but also extends the hydrophobicity of oligomer chains [26]. As molecules became thermally oligomerized, the longer oligomers migrated from the core to the outer interfaces in order to reduce the energy density gradient and grew highly ordered at the outer interfaces. Upon slow cooling, the order parameter of liquid crystal molecules increases, which can increase the number of defects at the boundary between nematic phases. This, in turn, reduces the γ [27,28]. This reduction in interfacial tension generates new interfaces, which initiates the geometrical transition from an isotropic sphere to fiber for micelles.

Until 6 h of thermal oligomerization, liquid crystalline micelles maintain a spherical shape due to the short $<\ell>$ (Figure S1a,b and Figure 2b–i). Spherical micelles are known to have 2–50 nm in diameter when the thermodynamic mechanism is dominant. Spherical micelles can further grow into larger lamellar structures. When the curvature is generated in a high aspect-ratio lamellar by kinetical fluctuation, various-sized micron-scale vesicles can be formed [29,30]. In our approach, liquid crystalline micelles are kinetic-driven vesicles considering the average diameter (d) of 12.37 ± 0.29 μm after 6 h of thermal oligomerization (Figure 2b-i and c). After 12 h of thermal oligomerization, spherical micelles changed into branched structures with a branch width (w) of 1.39 ± 0.25 μm and a larger diameter ($d = 62.19 \pm 44.37$ μm), as shown in Figure 2b-ii. Further thermal oligomerization resulted in a larger diameter ($d = 131.75 \pm 53.43$ μm) with a smaller fiber width ($w = 1.21 \pm 0.22$ μm) at 18 h oligomerization (Figure 2b-iii). After 24 h of thermal oligomerization, randomly coiled fibrous micelles were achieved with a greater diameter and a smaller fiber width ($d = 223.93 \pm 66.32$ μm, $w = 0.93 \pm 0.10$ μm) (Figure 2b-iv). Under a cross-polarized optical microscope, liquid crystalline oligomers demonstrate bright features in polarized optical microscope images, indicating nematic molecular alignments in both spherical and fibrous micelles (Figures S1 and S2). The oligomerization time-dependent growth of the liquid crystalline micelle is summarized in Figure 2c. The micelle's diameter increases with the fibrous part's decrement in width during the shape transition of liquid crystalline micelles from spherical to fibrous micelles under the thermal oligomerization process.

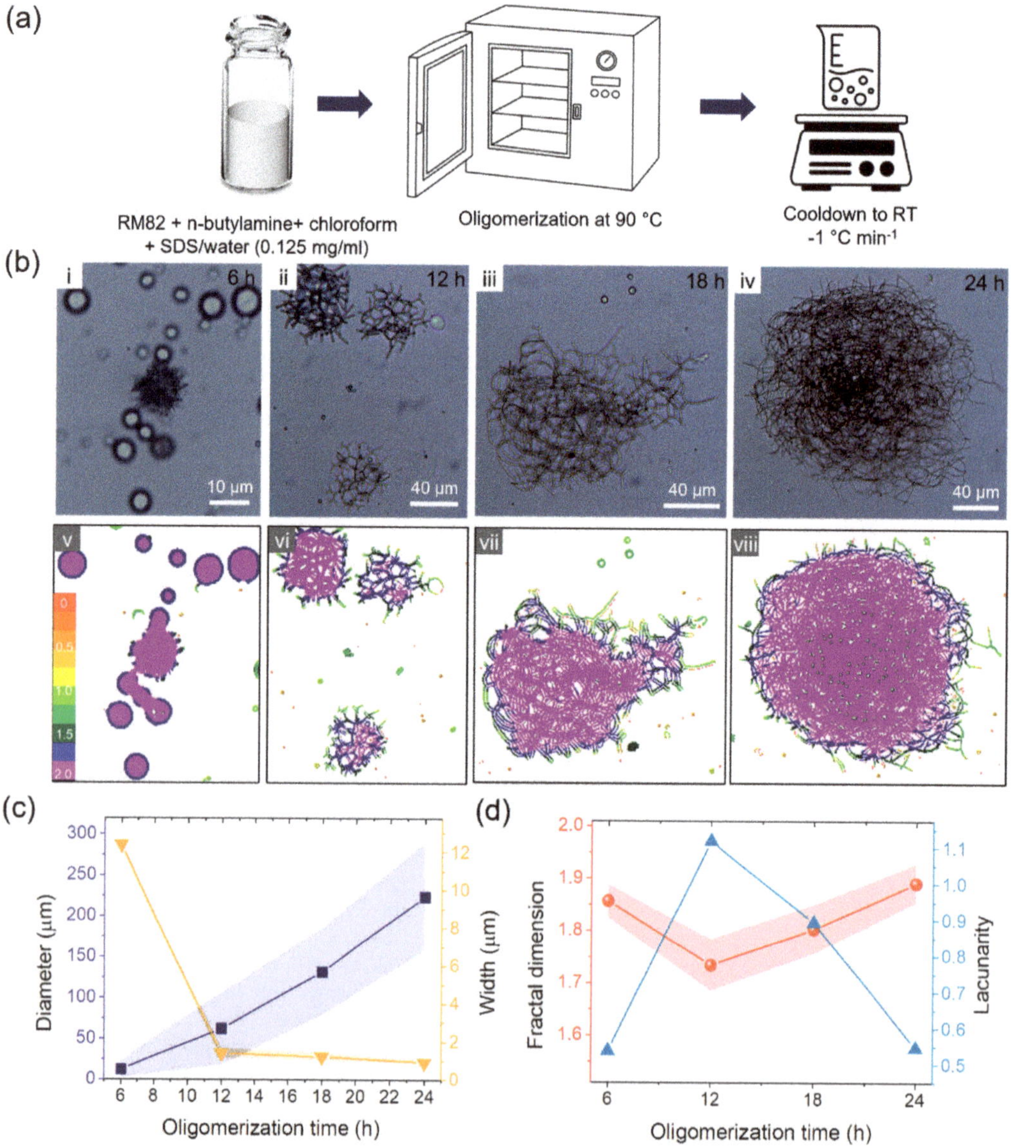

Figure 2. Shape transition of liquid crystalline oligomers by thermal oligomerization. (**a**) Schematic images of thermal oligomerization procedure. (**b**) Shape transition of liquid crystalline oligomers from spherical to branched and fibrous shapes visualized by optical microscopy (**i–iv**) and local connected fractal analysis (**v–viii**) against thermal oligomerization time of (**i**,**v**) 6 h, (**ii**,**vi**) 12 h, (**iii**,**vii**) 18 h, and (**iv**,**viii**) 24 h. (**c**) Diameter (■) and width (▼) of nematic liquid crystalline oligomers against thermal oligomerization time. (**d**) Fractal dimension (•) and lacunarity (▲) of nematic liquid crystalline oligomers against thermal oligomerization time.

For further quantification, locally connected fractal dimension (D_F) and lacunarity (λ) are analyzed, as shown in Figure 2d. A local complexity of patterns can be found in the locally connected fractal dimension. Its value displays the state of the dimension and how a pattern's scale affects the level of detail. The fractal value of a 2D picture is between 1 and 2, which indicates that the fractal is in a state halfway between a line and an area. The increasing fractal dimension shows the development of a complex pattern. The higher fractal dimension indicates the formation of a detailed pattern [31,32]. Lacunarity is an image's heterogeneity, which indicates its dispersity and homogeneity. For instance, the monodisperse sample has a low lacunarity value, and complicated pictures have a high lacunarity value [33]. The distribution of local connected fractal dimension was demonstrated in Figure 2b-v–viii, and results are plotted in Figure 2d. The fractal dimension decreased from 1.86 to 1.73 as the spherical micelle changed into the branched shape during 12 h oligomerization, which indicates increased linearity of the micelle. However, while micelles transition into fibrous shapes during further oligomerization, locally connected fractal dimension increased because entangled fibrous micelles merged into the area in 2D imagery. Lacunarity also increased during 12 h oligomerization due to the shape transition of the micelle and decreased as micelles changed in the fibrous branch.

To control the order of liquid crystalline monomers, directional ordering of liquid crystals in spatially confined condition must take place, as reported by Dong Ki Yoon and his coworkers [34]. Here, liquid crystalline micelles were thermally oligomerized in a micro-channel with a width of 50 μm, a depth of 20 μm, and a length of 7.5 mm (Figure S3) in the formation of spatial confinement during thermal oligomerization and shape transition. The microchannel was constructed with polydimethylsiloxane (PDMS) by soft lithography. The surface of the PDMS micro-channel was O_2 plasma-treated for surface cleaning and wetting of the liquid crystalline emulsion on the PDMS mold. After pouring the liquid crystalline emulsion into the PDMS micro-channel, a glass slide was placed on the micro-channel as a superstrate to prevent the evaporation of water during thermal oligomerization. Following thermal oligomerization, liquid crystalline micelles were photopolymerized by irradiation of 365 nm ultraviolet light at an intensity of 0.4 W cm^{-2} for 5 min (Figure 3a). Even in the micro-channel, spherical micelles were formed until 6 h of thermal oligomerization (Figure 3b-i). The spherical geometry of the micelle indicates that the micelle does not experience spatial confinement at this point, as the size of the micelle is still much smaller than the micro-channel. Hence, the size of spherical micelles was uniformly controlled. Within the spherical micelle, however, fibrous rough textures can be observed. A cross-polarized microscope image revealed that liquid crystalline oligomers have nematic molecular alignments, as shown in Figure 3b-ii–iv. After 12 h of thermal oligomerization, liquid crystalline oligomers were anisotropically entangled along the channel and in the intermediate state between spherical and fibrous shapes (Figure 3c). This morphological transformation became more obvious following further thermal oligomerization. After 24 h of thermal oligomerization reaction, the liquid crystalline oligomer constructed almost linear fibers within the micro-channel with a few discontinuous defects (Figure 3d). Finally, high-aspect-ratio linear fiber was constructed without any discernable defects after 25 h of thermal oligomerization (Figure 3e). Along fiber direction, liquid crystalline oligomers were aligned, as evidenced from the bright cross-polarized microscope image when the fiber had a 45° offset angle with respect to both the polarizer and the analyzer (Figure 3c,d,e-iii) and dark image at 0° (ii)/90° (iv) in Figure 3c–e.

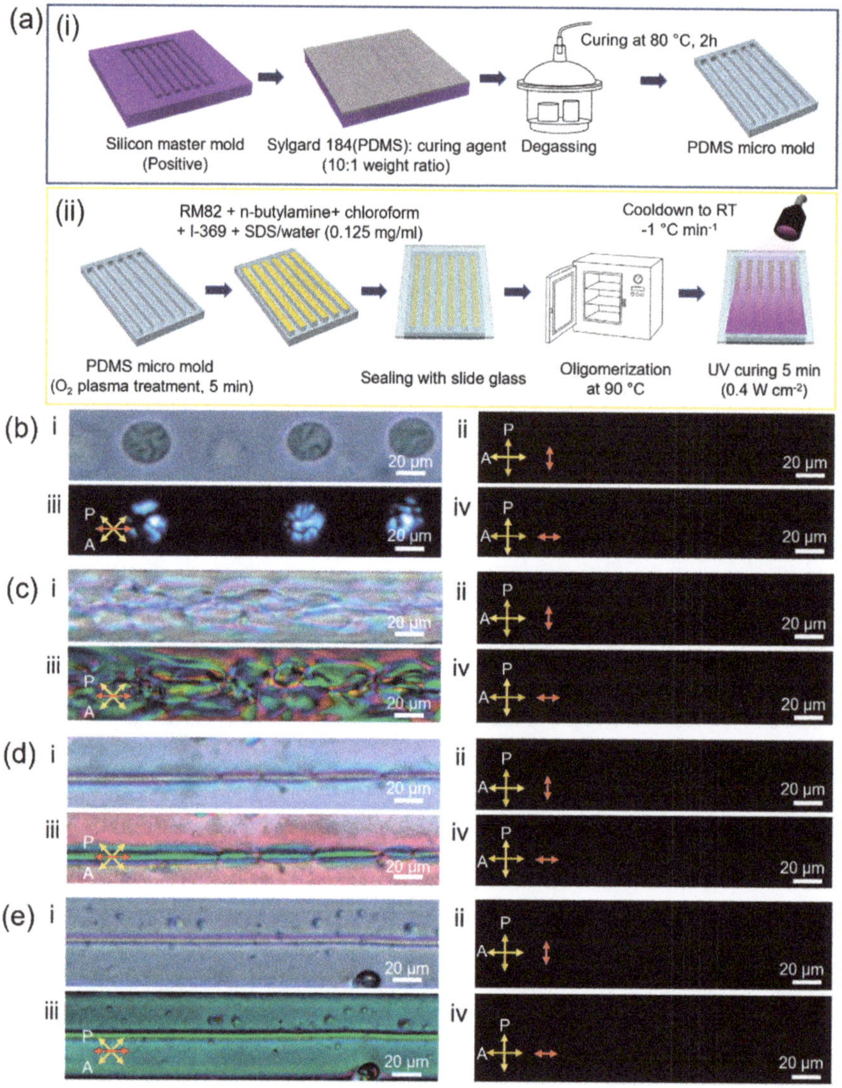

Figure 3. Spatial-confinement effects on shape changes in nematic liquid crystalline drops against thermal oligomerization times. (**a**) Schematic images of the construction of microchannel (**i**) and the thermal oligomerization procedure with spatial confinement (**ii**). (**b–e**) Shape transition of liquid crystalline oligomers from spherical to linear fiber geometry at different oligomerization times of (**b**) 6 h, (**c**) 12 h, (**d**) 24 h, and (**e**) 25 h. Scale bars are 20 μm. (**i**) Optical microscope image and (**ii–iv**) cross-polarized optical microscope images with a different directional offset of the micro-channel from the analyzer: (**ii**) 0°, (**iii**) 45°, and (**iv**) 90°. The red arrow indicates the long axis of the micro-channel.

In a case where no such spatial limitations exist, the size of the micelle gradually increases from 12.37 ± 0.29 μm to 223.93 ± 66.32 μm as thermal oligomerization time increases, as shown in Figure 2c. Here, as the self-assembly of liquid crystalline monomers was performed in limited confinements of a micro-channel, the width of the micro-channel sets the limit for micelle size. A micro-channel width of 50 μm limits the size of a micelle, producing spatial confinement effects along the width direction. This anisotropic confinement induces the growth of the liquid crystalline oligomer perpendicular to the width direction of the micro-channel. Consequently, a single 5 μm-thick liquid crystalline oligomer fiber could be achieved, shown in Figure 3e, as opposed to the construction of multiple-layered micelles in a limited space.

Figure 4a shows the Fast Fourier Transform (FFT) images of spatially confined liquid crystalline micelles at different thermal oligomerization times. To visually examine the alignment direction and degree of the pattern, images were transformed into the frequency domain and analyzed by the FFT method. The areas presented in the converted image depend on frequency size. If a pattern was aligned along a single direction, the FFT image would show a series of points perpendicular to the direction of the original image. On the other hand, if the image has a random pattern, a circular region is produced in the FFT image. As thermal oligomerization time increased, the micelle became more fibrous and molecules aligned, as evidenced from the brighter vertical line pattern. After 25 h of thermal oligomerization, a straight fiber-shaped micelle was constructed (Figure 3e). Its FFT image shows the brightest and clear vertical center line. The local fractal dimension is visualized in Figure 4b. Then, fractal dimension and lacunarity were analyzed in Figure 4c. After 6 h of thermal oligomerization, the spherical liquid crystalline oligomers had a fractal dimension of 1.82. After 12 h of thermal oligomerization, the liquid crystalline structures are in an intermediary state between spherical and fibrous shapes. Despite changes in shapes, the fractal dimension slightly decreased to 1.61. Meanwhile, lacunarity significantly increased due to the heterogeneity of the liquid crystalline structures. When thermal oligomerization time increased to 24 h, both local fractal dimension and lacunarity significantly decreased for fibers with only a few defects. Finally, the fractal dimension further decreased to 1.13 with a single fiber shape after 25 h of thermal oligomerization. As the fractal dimension of 1 corresponds to a perfect line, 1.13 of the local fractal dimensions indicates the linearity of the fiber. In addition to the smallest local fractal dimension, the smallest value of lacunarity indicates the uniformity of the fiber polymerized in the spatially confined channel.

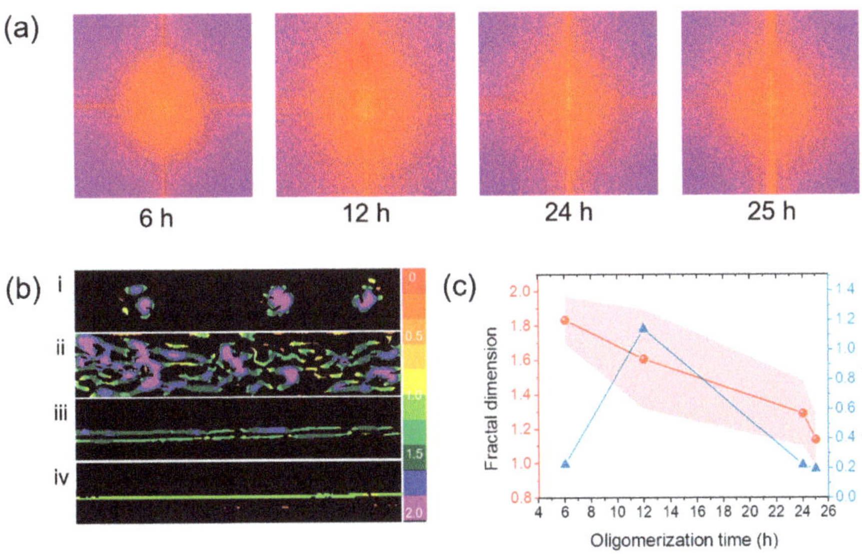

Figure 4. Spatial-confinement effects on shape change of nematic liquid crystal drop over oligomerization time. (**a**) Fast Fourier transform (FFT) images of spatially confined nematic liquid crystals at different oligomerization times. (**b**) Images of local connected fractal analysis at different oligomerization times: (i) 6 h, (**ii**) 12 h, (**iii**) 24 h, and (**iv**) 25 h. (c) Fractal dimension (•) and lacunarity (▲) of spatially confined grown nematic liquid crystals over oligomerization time.

In this study, we modified the architectures of liquid crystalline micelles from isotropic spherical to anisotropic fiber, which can reduce the surface tension by simply increasing thermal oligomerization time. When bottom-up molecular self-assembly was performed in a top-down micro-mold, we were able to manufacture single fibrous liquid crystal oligomers with a preferential molecular order along the micro-channel direction without the use of shears for wet-spinning or an electrical field for electrospinning.

3. Materials and Methods

In 13.5 mg of chloroform, 4.5 mg of 1,4-bis-[4-(6-acryloyloxyhexyloxy)benzoyloxy]-2-methylbenzene (RM82, Synthon Chemicals, Wolfen, Germany) was dissolved with 0.5 mg of n-butylamine (Butylamine, Sigma Aldrich, St. Louis, MO, USA) and 2-benzyl-2-dimethylamino-1-(4-morpholinophenyl)-butanone-1 (I-369, Ciba, Basel, Switzerland). The liquid crystalline solution was mixed with an aqueous surfactant solution of 0.125 mg/mL of sodium dodecyl sulfate (SDS, Sigma Aldrich). Resultant suspensions were placed in a convection oven at 90 °C for thermal oligomerization for 6, 12, 18, and 24 h to observe the growth of liquid crystalline oligomers.

For the spatially confined growth experiment, a channel-shaped polydimethylsiloxane (PDMS) micro-mold was prepared with the following method. The PDMS precursor (Sylgard 184, Dow Corning, Midland, MI, USA) was mixed with a curing agent at a 10:1 (PDMS precursor: curing agent) weight ratio. The PDMS mixture was poured onto a positive silicon master mold featured with microchannels (50 μm width, 20 μm depth, and 7.5 mm length). The PDMS negative mold was prepared by placing it in a vacuum oven for thermal curing at 80 °C for 2 h. The PDMS microchannel was detached from the silicon mold after curing. The PDMS microchannel and slide glass were treated with air plasma for 5 min for surface cleaning. The liquid crystalline emulsion was poured into PDMS microchannels and sealed with a slide glass superstrate. The emulsion was thermally oligomerized in a 90 °C vacuum oven for 6, 12, 24, and 25 h. After thermal oligomerization, liquid crystalline oligomers were photo-polymerized upon exposure to 365 nm ultraviolet light at an intensity of 0.4 W cm^{-2} for 10 min. The photopolymerized liquid crystalline polymers were observed by a cross-polarized optical microscope (Nikon eclipse, Nikon, Tokyo, Japan) and analyzed by ImageJ and FracLac plugins.

4. Conclusions

In conclusion, we successfully polymerized liquid crystalline polymers in high-aspect-ratio fiber geometry by a hybrid approach. During the thermal oligomerization of liquid crystalline monomers, bottom-up molecular self-assembly is achieved within a micro-channel prepared by a top-down approach, followed by photopolymerization. Although most fiber production processes employ a high shear rate in a spinning process and high electrical voltage in an electrospinning process, this work suggests that spatial confinement can induce the self-assembly of directional fibrous micelle without applying large shear forces during the manufacturing process. Within high-aspect-ratio microfiber, we were able to generate nematic molecular alignments of the liquid crystalline polymers along the long axis of the fiber. This approach has potential in confinement-driven liquid crystalline colloidal systems for applications in non-conventional high-aspect-ratio fibers as well as soft actuators.

Supplementary Materials: The following supporting information can be downloaded at: https://www.mdpi.com/article/10.3390/molecules27175639/s1; Figure S1: Shape transition of liquid crystal oligomers from spherical to flower-shaped micelle (red arrow) over oligomerization time; Figure S2: Transmission optical microscope images and cross-polarized optical microscope image; Figure S3: Schematic images of micro-channel with dimensional information for length and depth of the micro-channel, optical microscope images and cross-polarized optical microscope image of the micro-channel.

Author Contributions: Conceptualization, J.J.W.; methodology, J.G.K.; validation, J.G.K. and J.G.L.; formal analysis, J.G.K.; investigation, J.G.K. and J.G.L.; resources, J.J.W.; data curation, J.G.K.; writing—original draft preparation, J.G.K. and J.J.W.; writing—review and editing, J.J.W.; visualization, J.G.K.; supervision, J.J.W.; project administration, J.J.W.; funding acquisition, J.J.W. All authors have read and agreed to the published version of the manuscript.

Funding: This research was funded by Inha University, 63092-1.

Institutional Review Board Statement: Not applicable.

Informed Consent Statement: Not applicable.

Data Availability Statement: Data are contained within the article.

Conflicts of Interest: The authors declare no conflict of interest.

Sample Availability: Samples of the compounds are not available from the authors.

References

1. Kim, J.G.; Park, J.E.; Won, S.; Jeon, J.; Wie, J.J. Contactless Manipulation of Soft Robots. *Materials (Basel)* **2019**, *12*, 3065. [CrossRef] [PubMed]
2. Tchamba Yimga, N.; Ramanan, C.; Borchert, H.; Parisi, J.; Untenecker, H.; Kirsch, P.; Von Hauff, E. Interplay between Long-Range Crystal Order and Short-Range Molecular Interactions Tunes Carrier Mobility in Liquid Crystal Dyes. *ACS Appl. Mater. Interfaces* **2017**, *9*, 6228–6236. [CrossRef] [PubMed]
3. Ula, S.W.; Traugutt, N.A.; Volpe, R.H.; Patel, R.R.; Yu, K.; Yakacki, C.M. Liquid Crystal Elastomers: An Introduction and Review of Emerging Technologies. *Liq. Cryst. Rev.* **2018**, *6*, 78–107. [CrossRef]
4. Gelebart, A.H.; Jan Mulder, D.; Varga, M.; Konya, A.; Vantomme, G.; Meijer, E.W.; Selinger, R.L.B.; Broer, D.J. Making Waves in a Photoactive Polymer Film. *Nature* **2017**, *546*, 632–636. [CrossRef]
5. Ohm, C.; Brehmer, M.; Zentel, R. Liquid Crystalline Elastomers as Actuators and Sensors. *Adv. Mater.* **2010**, *22*, 3366–3387. [CrossRef]
6. Ge, F.; Zhao, Y. Microstructured Actuation of Liquid Crystal Polymer Networks. *Adv. Funct. Mater.* **2020**, *30*, 1901890. [CrossRef]
7. Kularatne, R.S.; Kim, H.; Boothby, J.M.; Ware, T.H. Liquid Crystal Elastomer Actuators: Synthesis, Alignment, and Applications. *J. Polym. Sci. Part B Polym. Phys.* **2017**, *55*, 395–411. [CrossRef]
8. Pang, X.; Lv, J.-A.; Zhu, C.; Qin, L.; Yu, Y. Photodeformable Azobenzene-Containing Liquid Crystal Polymers and Soft Actuators. *Adv. Mater.* **2019**, *31*, 1904224. [CrossRef]
9. White, T.J.; Broer, D.J. Programmable and Adaptive Mechanics with Liquid Crystal Polymer Networks and Elastomers. *Nat. Mater.* **2015**, *14*, 1087–1098. [CrossRef]
10. Ahn, S.K.; Ware, T.H.; Lee, K.M.; Tondiglia, V.P.; White, T.J. Photoinduced Topographical Feature Development in Blueprinted Azobenzene-Functionalized Liquid Crystalline Elastomers. *Adv. Funct. Mater.* **2016**, *26*, 5819–5826. [CrossRef]
11. Wie, J.J.; Shankar, M.R.; White, T.J. Photomotility of Polymers. *Nat. Commun.* **2016**, *7*, 13260. [CrossRef] [PubMed]
12. Ube, T.; Kawasaki, K.; Ikeda, T. Photomobile Liquid-Crystalline Elastomers with Rearrangeable Networks. *Adv. Mater.* **2016**, *28*, 8212–8217. [CrossRef] [PubMed]
13. Camacho-Lopez, M.; Finkelmann, H.; Palffy-Muhoray, P.; Shelley, M. Fast Liquid-Crystal Elastomer Swims into the Dark. *Nat. Mater.* **2004**, *3*, 307–310. [CrossRef]
14. Van Oosten, C.L.; Bastiaansen, C.W.M.; Broer, D.J. Printed Artificial Cilia from Liquid-Crystal Network Actuators Modularly Driven by Light. *Nat. Mater.* **2009**, *8*, 677–682. [CrossRef] [PubMed]
15. Yue, Y.; Norikane, Y.; Azumi, R.; Koyama, E. Light-Induced Mechanical Response in Crosslinked Liquid-Crystalline Polymers with Photoswitchable Glass Transition Temperatures. *Nat. Commun.* **2018**, *9*, 3234. [CrossRef]
16. Jeon, J.; Choi, J.C.; Lee, H.; Cho, W.; Lee, K.; Kim, J.G.; Lee, J.W.; Joo, K.-I.; Cho, M.; Kim, H.R.; et al. Continuous and Programmable Photomechanical Jumping of Polymer Monoliths. *Mater. Today* **2021**, *49*, 97–106. [CrossRef]
17. Tian, H.; Wang, Z.; Chen, Y.; Shao, J.; Gao, T.; Cai, S. Polydopamine-Coated Main-Chain Liquid Crystal Elastomer as Optically Driven Artificial Muscle. *ACS Appl. Mater. Interfaces* **2018**, *10*, 8307–8316. [CrossRef]
18. Kim, J.G.; Jeon, J.; Sivakumar, R.; Lee, J.; Kim, Y.H.; Cho, M.; Youk, J.H.; Wie, J.J. Light-Fueled Climbing of Monolithic Torsional Soft Robots via Molecular Engineering. *Adv. Intell. Syst.* **2022**, *4*, 2100148. [CrossRef]
19. Liu, X.; Xu, Y.; Heuts, J.P.A.; Debije, M.G.; Schenning, A.P.H.J. Monodisperse Liquid Crystal Network Particles Synthesized via Precipitation Polymerization. *Macromolecules* **2019**, *52*, 8339–8345. [CrossRef]
20. Park, S.; Lee, S.S.; Kim, S.H. Photonic Multishells Composed of Cholesteric Liquid Crystals Designed by Controlled Phase Separation in Emulsion Drops. *Adv. Mater.* **2020**, *32*, 2002166. [CrossRef]
21. Gwon, S.; Park, S. Preparation of Uniformly Sized Interpenetrating Polymer Network Polyelectrolyte Hydrogel Droplets from a Solid-State Liquid Crystal Shell. *J. Ind. Eng. Chem.* **2021**, *99*, 235–245. [CrossRef]
22. Lee, J.Y.; Oh, M.J.; Yoo, P.J. Concentric/Bipolar Ordering of Liquid Crystalline Graphene Oxide Nanosheets Confined in Microfluidically Synthesized Spherical Droplets. *J. Mater. Chem. C* **2021**, *9*, 8947–8954. [CrossRef]

23. Wei, W.S.; Xia, Y.; Ettinger, S.; Yang, S.; Yodh, A.G. Molecular Heterogeneity Drives Reconfigurable Nematic Liquid Crystal Drops. *Nature* **2019**, *576*, 433–436. [CrossRef] [PubMed]
24. Ware, T.H.; McConney, M.E.; Wie, J.J.; Tondiglia, V.P.; White, T.J. Voxelated Liquid Crystal Elastomers. *Science* **2015**, *347*, 982–984. [CrossRef] [PubMed]
25. Gruler, H. The Elastic Constants of a Nematic Liquid Crystal. *Z. Naturforsch. -Sect. A J. Phys. Sci.* **1975**, *30*, 230–234. [CrossRef]
26. Tintaru, M.; Moldovan, R.; Beica, T.; Frunza, S. Surface Tension of Some Liquid Crystals in the Cyanobiphenyl Series. *Liq. Cryst.* **2001**, *28*, 793–797. [CrossRef]
27. Guttman, S.; Sapir, Z.; Schultz, M.; Butenko, A.V.; Ocko, B.M.; Deutsch, M.; Sloutskin, E. How Faceted Liquid Droplets Grow Tails. *Proc. Natl. Acad. Sci. USA* **2016**, *113*, 493–496. [CrossRef]
28. Gannon, M.G.J.; Faber, T.E. The Surface Tension of Nematic Liquid Crystals. *Philos. Mag. A* **1978**, *37*, 117–135. [CrossRef]
29. Lu, Y.; Zhang, E.; Yang, J.; Cao, Z. Strategies to Improve Micelle Stability for Drug Delivery. *Nano Res.* **2018**, *11*, 4985–4998. [CrossRef]
30. Stuart, M.C.A.; Boekema, E.J. Two Distinct Mechanisms of Vesicle-to-Micelle and Micelle-to-Vesicle Transition Are Mediated by the Packing Parameter of Phospholipid-Detergent Systems. *Biochim. Biophys. Acta Biomembr.* **2007**, *1768*, 2681–2689. [CrossRef]
31. Shin, J.I.; Cho, S.J.; Jeon, J.; Lee, K.H.; Wie, J.J. Three-Dimensional Micropatterning of Semiconducting Polymers: Via Capillary Force-Assisted Evaporative Self-Assembly. *Soft Matter* **2019**, *15*, 3854–3863. [CrossRef] [PubMed]
32. Xia, Y.; Cai, J.; Perfect, E.; Wei, W.; Zhang, Q.; Meng, Q. Fractal Dimension, Lacunarity and Succolarity Analyses on CT Images of Reservoir Rocks for Permeability Prediction. *J. Hydrol.* **2019**, *579*, 124198. [CrossRef]
33. Plotnick, R.E.; Gardner, R.H.; O'Neill, R.V. Lacunarity Indices as Measures of Landscape Texture. *Landsc. Ecol.* **1993**, *8*, 201–211. [CrossRef]
34. Ryu, S.H.; Ahn, H.; Shin, T.J.; Yoon, D.K. Direct Observation of Liquid Crystal Phases under Nanoconfinement: A Grazing Incidence X-Ray Diffraction Study. *Liq. Cryst.* **2017**, *44*, 713–721. [CrossRef]

molecules

MDPI

Communication

Polymer Dispersed Liquid Crystal Imprinted by Microlens Array for Enhanced Outcoupling Efficiency of Organic Light Emitting Diode

Seongmin Lim [1], Hyeon-Sik Ahn [1], Eun-Jeong Jang [1], So-Young Boo [2], Akpeko Gasonoo [3], Jin-Seog Gwag [4], Jae-Hyun Lee [2] and Yoonseuk Choi [1,*]

[1] Department of Electronic Engineering, Hanbat National University, Daejeon 34158, Republic of Korea; seongmin625@naver.com (S.L.); princass123@naver.com (H.-S.A.); enjeong722@gmail.com (E.-J.J.)

[2] Department of Creative Convergence Engineering, Hanbat National University, Daejeon 34158, Republic of Korea; mick0327@naver.com (S.-Y.B.); jhyunlee@hanbat.ac.kr (J.-H.L.)

[3] Department of Chemistry, University of Calgary, Calgary, AB T2N 1N4, Canada; amoebatheo2009@gmail.com

[4] Department of Physics, Yeungnam University, Gyeongsan 38541, Republic of Korea; sweat3000@ynu.ac.kr

* Correspondence: ychoi@hanbat.ac.kr; Tel.: +82-42-821-1134

Abstract: In this paper, we demonstrate the use of polymer dispersed liquid crystal (PDLC) imprinted with a microlens array (MLA) via solution process to improve the outcoupling efficiency of organic light emitting diodes (OLEDs). The PDLC, well known for its scattering effect, is an excellent technology for improving the outcoupling efficiency of OLEDs. Additionally, we introduce a simple spin-coating process to fabricate PDLC which is adaptable for future solution-processed OLEDs. The MLA-imprinted PDLC applied OLED shows an enhancement factor of 1.22 in outcoupling efficiency which is a 37.5% increase compared to the existing PDLC techniques without changing the electrical properties of the OLED. Through this approach, we can expect the roll-to-roll based extremely flexible OLED, and with further research on pattering PDLC by various templates, higher outcoupling efficiency is achievable through a simple UV irradiation process.

Keywords: polymer dispersed liquid crystal; outcoupling efficiency; organic light emitting diode

1. Introduction

Organic light emitting diodes (OLEDs) are flexible electronic devices renowned for their thinness, compact size, and bendability, along with exceptional color reproduction, making them prevalent in today's display industry [1,2]. OLEDs consume less power compared to light emitting diodes and liquid crystal display technologies. Due to their rapid response time and superior contrast ratio, they are being extensively researched as a next generation display technology [3,4]. An OLED comprises multiple layers of organic materials stacked between two electrodes, the anode and the cathode situated above and below the substrate. However, a significant amount of the light emitted is absorbed at the interface between the OLED device and the air, as well as within the multi-layer organic material due to total internal reflection caused by the large critical angle stemming from the refractive index disparity between layers. Consequently, only about 20% of the light is emitted externally, leading to low efficiency [5–7]. To combat this, numerous studies aim to enhance the external light extraction efficiency of OLED devices to improve power efficiency and extend their lifespan [8,9]. Light extraction technologies primarily focus on minimizing light loss by reducing the refractive index mismatch within the OLED, thereby lessening total internal reflection [10,11]. Various optical structures have been integrated into OLED devices for this purpose, including microlens array (MLA) [12,13], visible parylene film [14], MLA patterned parylene substrate [15], random surfaces [16], porous cellulose paper [17], and light scattering layers [18–20]. Among these studies, research using parylene has produced a light extraction film that can serve as a flexible substrate and exhibits a high outcoupling efficiency exceeding 20%. However, the drawback is that this film is manufactured using a high temperature and vacuum based chemical vapor deposition (CVD) process, necessitating the establishment of a specific process environment, and resulting in long production times and high costs. Research involving random-surface-based light extraction relies on plasma and gas, while cellulose-based research requires chemical treatments like potassium hydroxide (KOH) and sodium chloride (NaCl), along with extended chemical reaction times for cellulose purification. In both cases, the time required for light extraction film production is lengthy, and the process is complex. In studies using a light scattering layer, a light extraction film using PDLC has been created, but it cannot be applied to flexible OLEDs because it utilizes two glass substrates. Consequently, there is a need for an external light extraction film process with a simplified production environment and process, capable of significantly reducing production time and cost, and suitable for application to flexible OLEDs. Traditional external PDLC light extraction films, which have been investigated for OLED efficiency, are unsuitable for flexible OLEDs due to their reliance on rigid glass substrates and a thick PDLC layer. When the PDLC carries an MLA pattern, it can further improve the outcoupling efficiency of the OLED. Furthermore, MLA-imprinted PDLC prepared by a solution process is expected to enable the production of large area flexible outcoupling films by adopting roll-to-roll, blade, and bar coating methods.

In this paper, our objective is to enhance the outcoupling efficiency of OLEDs by creating a polymer dispersed liquid crystal (PDLC) imprinted with an MLA pattern using spin coating, a solution process method. The use of PDLC for external light extraction has been a long-studied approach to augment OLED performance [21]. PDLC consists of a polymer matrix with liquid crystal (LC) droplets that scatter incoming light by dispersing liquid crystal droplets of several microns in size throughout the polymer matrix. The fabrication process of the proposed MLA-imprinted PDLC is straightforward, involving only spin coating and UV irradiation without the need for any pretreatment to detach the MLA substrate from the PDLC. MLA-imprinted PDLCs exhibit 41% higher haze than conventional PDLCs and have a better light scattering effect. The MLA-imprinted PDLC applied OLED shows an enhancement factor of 1.22 in outcoupling efficiency which is a 37.5% increase compared to the existing PDLC techniques without changing the electrical properties of OLED [22]. MLA-imprinted PDLC is flexible and has high outcoupling efficiency, which is expected to improve the outcoupling efficiency of OLEDs and be used as a flexible substrate.

2. Results and Discussion

This PDLC is a liquid-crystal-based material that has long been studied for its light scattering effect to increase the outcoupling efficiency of OLED [23]. Haze is crucial in achieving this effect: the greater the haze, the more the light scatters. As light emitted from the OLED passes through the PDLC, it scatters in various directions. Subsequently, the refractive index at different interfaces changes, allowing part of the light to be redirected and spread outward, which enhances the outcoupling efficiency [24]. The MLA also contributes to improved outcoupling efficiency by refracting incident light at the air lens interface, altering its path and influencing optical properties such as the light outcoupling effect, intensity, and angular distribution [25]. Our research investigates the enhancement of OLED outcoupling efficiency by combining these two methods. To assess the optical impact of the PDLC imprinted with the MLA pattern, a PDLC layer without a pattern was produced for comparison. The resulting PDLC layer had a thickness of 18 μm, while the planarization layer used to flatten the surface formed a thin film of 36.6 μm.

Figure 1 shows an actual photograph of MLA-imprinted PDLC. Figure 1a shows that the PDLC solution has good coating properties, and the edges of the film are where the solution was subjected to surface tension during spin coating. Figure 1b shows the

MLA-imprinted PDLC when bent. MLA-imprinted PDLC is flexible and resilient, so it can be expected to be used as a substrate for flexible OLED.

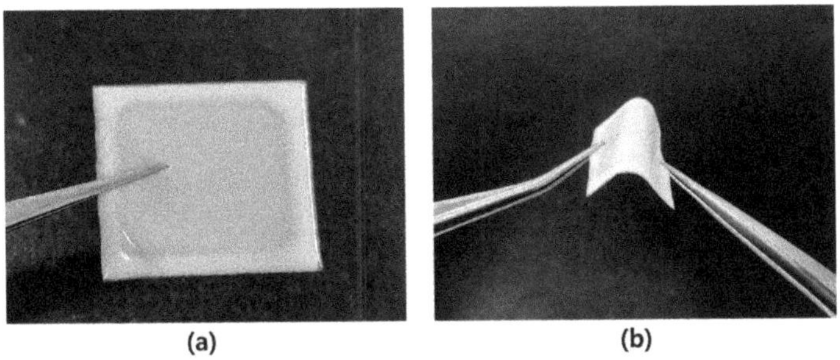

Figure 1. Actual photography of (**a**) MLA-imprinted PDLC; (**b**) bended MLA-imprinted PDLC.

Surface analysis was conducted using SEM to verify the successful imprint of the MLA pattern on the PDLC. Figure 2a shows the pattern of the MLA substrate used, while Figure 2b shows the PDLC with the MLA pattern imprinted after being peeled from the MLA. The MLA substrate utilized had a diameter of 20 μm and a height of 6 μm, and the MLA-imprinted PDLC replicated these dimensions with a diameter of 20 μm but a reduced height of 4 μm. The 2 μm discrepancy in height is attributed to the high viscosity of the NOA 63 solution, which has a viscosity of 2000 cps, preventing the PDLC solution from fully penetrating between the MLA patterns, though the diameter remained consistent. To address this, using a polymer matrix with lower viscosity or heating the solution before coating may aid in the pattern imprinting. The effectiveness of the MLA patterning process was confirmed as it maintained excellent pattern formation even after the peeling process, indicating that the solution process method can achieve impressive MLA patterns, akin to those produced by the deposition process.

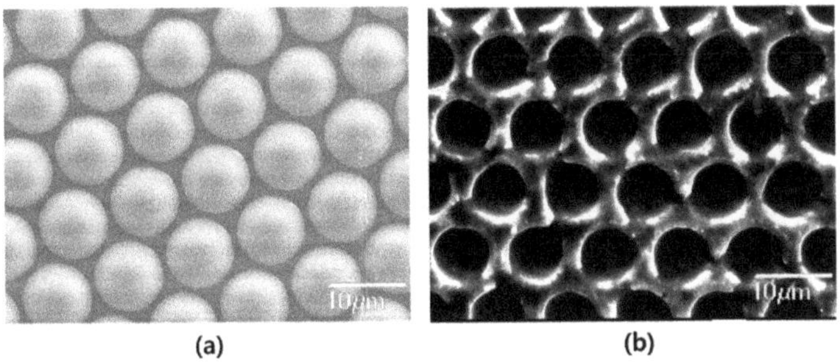

Figure 2. Scanning electron microscopy (SEM) images of: (**a**) the MLA substrate; (**b**) the MLA-imprinted PDLC.

Optical measurements were taken to assess the optical properties of the two types of PDLC. Figure 3a indicates the total transmittance (Tt) and parallel transmittance (Tp) of both the MLA-imprinted and non-imprinted PDLC, while Figure 3b indicates the haze for each type of PDLC. In Figure 3a, a decrease in transmittance with wavelength is observed, which is due to the change in refractive index with the frequency of light, resulting in wavelength-dependent light scattering or absorption. Consequently, as depicted in Figure 3b, the haze increases with shorter wavelengths. At a wavelength of 530 nm, the MLA-imprinted PDLC exhibited a Tt of 58.9% and a Tp of 0.9%, whereas the non-imprinted PDLC showed a Tt of 65% and a Tp of 4.5%. The lower Tp of the MLA-imprinted PDLC indicates a relatively higher light scattering effect compared to the non-imprinted PDLC. When incident light is scattered, Tp decreases and haze increases. The haze value was calculated using a formula that combines Tt and Tp [26]. At 530 nm, the non-imprinted PDLC had a haze of 93%, and the MLA-imprinted PDLC had a haze of 98.4%. With its higher haze value, the MLA-imprinted PDLC has the potential to enhance the outcoupling efficiency of OLEDs. Therefore, the efficiency improvement of OLEDs using MLA-imprinted PDLC was investigated.

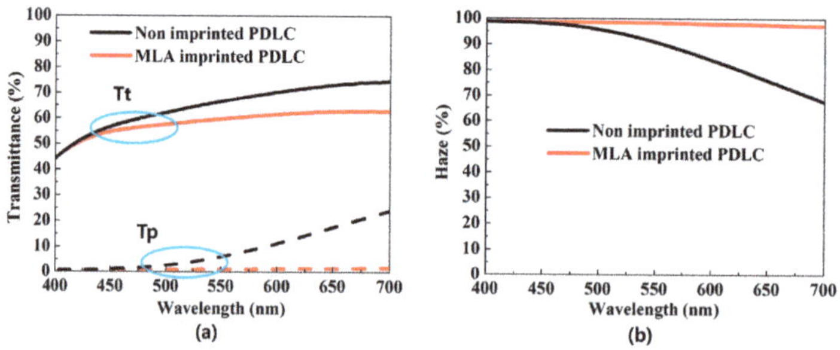

Figure 3. (**a**) Total transmittance and parallel transmittance; (**b**) haze of PDLC without MLA imprint and MLA-imprinted PDLC.

Figure 4 indicates the electrical and optical characteristics of the reference OLED and the OLED with MLA-imprinted PDLC. The solid symbol is current density-voltage, empty symbol is voltage-luminance in Figure 4a indicates the J-V-L characteristics of both the reference OLED and the OLED with MLA-imprinted PDLC, measured from 1 to 6 V. Both devices were activated at 3 V and exhibited similar J-V characteristics; however, the MLA-imprinted PDLC achieved higher luminance at the same voltage. The reference OLED reached a peak of 3632 cd/m^2, while the MLA-imprinted PDLC attained a maximum of 4792 cd/m^2, indicating an increase in luminance efficiency. Figure 4b indicates the EQE of each device. EQE is calculated by multiplying the internal quantum efficiency by the outcoupling efficiency, which allows for an analysis of the ratio of generated photons to injected electrons. As depicted in the EQE graph, a higher luminous efficiency is achieved relative to the number of electrons injected. The reference OLED had a maximum EQE of 6.1%, and this increased to 7.9% when the MLA-imprinted PDLC was applied, indicating an enhancement in EQE. Therefore, it can be concluded that the MLA-imprinted PDLC does not impair the electrical characteristics of the OLED. Instead, it offers improved luminance and power efficiency relative to the reference OLED driving voltage. Additionally, due to its superior luminous performance compared to current efficiency, the MLA-imprinted PDLC is anticipated to be a viable candidate for use as an external light extraction film for OLED.

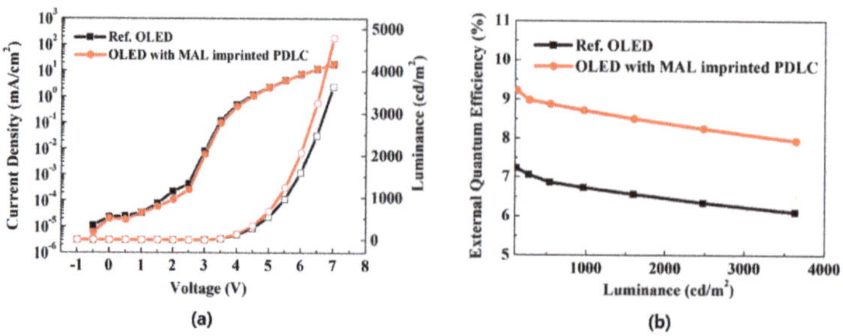

Figure 4. (**a**) Current density –voltage –luminance (J – V – L); (**b**) external quantum efficiency (EQE) of reference OLED and OLED with MLA-imprinted PDLC film.

The analysis of the impact of applying MLA-imprinted PDLC to OLEDs on outcoupling efficiency is presented in Figure 5. Figure 5a indicates the electroluminescence spectrum (EL) of the reference OLED and the OLED with MLA-imprinted PDLC operating at 5 V. It can also be seen that there is a double peak. This is because the phosphorescent dopant, bis(2-phenylpyridine)(acetylacetonate)iridum(III) ($Ir(ppy)_2(acac)$), consists of iridium, the central atom of which is a heavy atom. The heavy atom effect and the vibronic effect play an important role in controlling the properties of light emitting devices. One of the phenomena caused by these two effects is shoulder peek [27]. Figure 5b indicates a photograph comparing an OLED with the reference OLED and one with MLA-imprinted PDLC applied. A pixel with a size of 4 mm^2 was produced on the electrode. The reference OLED exhibited an EL intensity of up to 0.36%, while the OLED with MLA-imprinted PDLC showed an increased EL intensity of up to 0.44%. This analysis of outcoupling efficiency improvement confirms that the MLA-imprinted PDLC has an enhancement factor of 1.22. This enhancement is attributed to the strong scattering effect of the MLA-imprinted PDLC and the reduction in total internal reflection as the incident angle of light is decreased below the critical angle at the interface of the light extraction film and air. In summary, the results confirm that MLA-imprinted PDLC effectively improves the outcoupling efficiency of OLED compared to the existing PDLC.

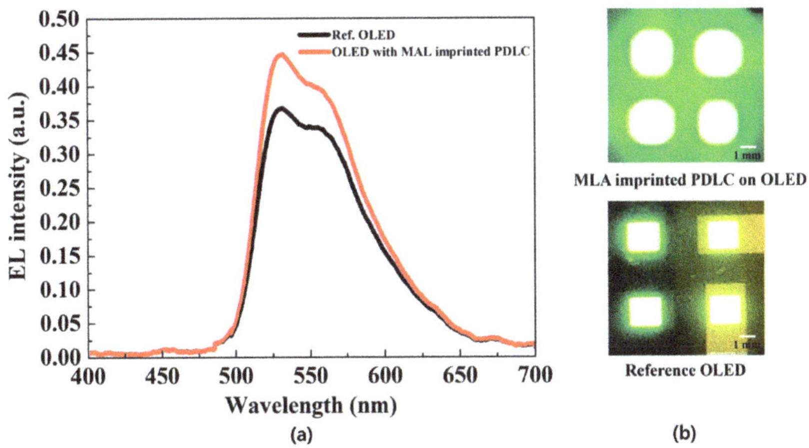

Figure 5. (**a**) Electroluminescence spectrum; (**b**) photographs of OLED with MLA-imprinted PDLC (**top**) and reference OLED (**bottom**).

3. Materials and Methods

Figure 6a illustrates the manufacturing process of MLA-imprinted PDLC using a spin-coating approach. For the preparation of PDLC solution, nematic liquid crystal (E7, Qingdao QY Liquid Crystal Co., Ltd., Shandong, China) with refractive index $n = 1.52$ and UV curing adhesive (NOA 63, Norland Product, Jamesburg, NJ, USA) a polymer matrix with refractive index $n_p = 1.52$, were used to match the refractive index of the used materials to ensure a smooth gap between the OLED and the interface. The difference in refractive index was minimized. This material was mixed at a 40:60 ratio. To ensure smooth mixing of liquid crystal and NOA 63, mixing was carried out at 80 °C, which is the phase transition temperature of liquid crystal.

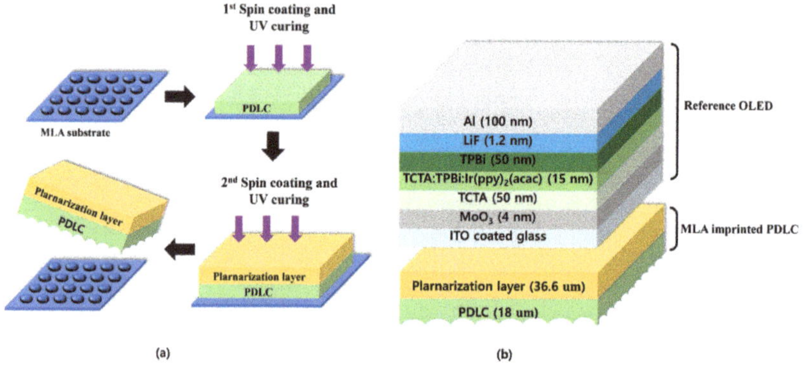

Figure 6. (**a**) Fabrication process of MLA-imprinted PDLC; (**b**) structure of the reference bottom-emitting OLED and the OLED with MLA-imprinted PDLC.

Next, the MLA substrate was precoated with the PDLC solution at 500 rpm for 5 s and then coated at 1500 rpm for 20 s. UV curing was performed in a UV oven at 150 °C for 15 min at an intensity of 32 mW/cm^2. This process formed a thin PDLC film on the MLA substrate, resulting in MLA-imprinted PDLC. For films without a pattern, a PDLC thin film was formed under the same conditions on a PET substrate. It is critical to note that the MLA-imprinted PDLC film formed after spin coating is very thin. Post spin coating, surface curvature may occur in the PDLC due to the MLA pattern, leading to an uneven film. If applied to OLED devices in this state, discrepancies in refractive index could produce negative effects. Hence, post imprinting planarization of the film is necessary. NOA 63 is used for its adjustable thickness and transparency, minimizing the impact on OLED light. The planarization spin-coating process involved precoating at 500 rpm for 5 s, followed by coating at 2000 rpm for 20 s. UV curing for 5 min then formed a thin film, which was easily peeled from the MLA substrate without pretreatment, readying a film applicable to OLED.

Figure 6b illustrates the structure of the bottom-emitting reference OLED and the positions where the MLA-imprinted PDLC is applied to comparatively analyze the optical and outcoupling efficiency of the fabricated MLA-imprinted PDLC. The glass substrate, coated with indium tin oxide (ITO, 150 nm), was cleaned by soaking in acetone, isopropyl alcohol, or boiled isopropyl alcohol using an ultrasonic cleaner and then dried in a vacuum oven at 150 °C for 15 min to prepare it for OLED deposition. The OLED organic layers were deposited sequentially as follows: molybdenum oxide (MoO_3, 4 nm) as a hole injection layer on the precleaned ITO coated glass, 4,4′,4′′-tris(carbazol-9-yl)triphenylamine (TCTA, 50 nm) as a hole transport material, bis(2-phenylpyridine)(acetylacetonate)iridium(III) ($Ir(ppy)_2(acac)$, 15 nm) as a phosphorescent dopant, 1,3,5-tris(N-phenylbenzimidizol-2-yl)benzene (TPBi, 50 nm) as an electron transport layer, and lithium fluoride (LiF, 12 nm) as an electron injection layer. Finally, aluminum (Al, 100 nm) was used as the cathode material.

An alpha-step device (KLA, D-600) was used to measure the thickness of the MLA-imprinted PDLC produced through a solution process using spin coating. Additionally, the surface of the PDLC was analyzed using a scanning electron microscope (SEM, JEOL) to confirm the quality of the MLA imprinting. The optical properties, including transmittance and haze, of the produced MLA-imprinted PDLC and non-imprinted PDLC, were analyzed in the wavelength range of 400 to 700 nm using a UV-visible-NIR spectrometer (PerkinElmer, Lambda 950). To evaluate the efficiency improvement by applying MLA-imprinted PDLC to OLEDs, the current density-voltage-luminance (J-V-L) characteristics and external quantum efficiency (EQE) were measured and analyzed using a source meter (KEITHLEY 2400) and analyzed in the vertical direction using Photo Research (LMS PR 650) software. Voltage measurements were performed in the range of 1 V to 7 V at 0.5 V intervals. The analysis of the outcoupling efficiency before and after the application of MLA-imprinted PDLC was performed using an integrated sphere (IS200-4) and a spectrometer (Thorlabs, CCS200/M).

4. Conclusions

In this paper, a PDLC imprinted with an MLA is produced via a solution process to enhance the outcoupling efficiency of OLEDs. This process involving spin coating and UV irradiation can facilitate low-temperature and rather simple fabrication which effectively reduces manufacturing costs and time. Also, this novel PDLC film is flexible and very thin which is an advantage for adopting future full flexible OLEDs without changing any other electrical characteristics of the OLED since it is an external light extraction method. The haze of fabricated PDLC film with MLA pattern was 98.4% which is higher than previous PDLC approaches and resulted in increased outcoupling efficiency. The enhancement factor of outcoupling efficiency with this film was 1.22 which is a 37.5% increase compared to the previous PDLC-based OLED. Furthermore, this PDLC method can vary the spatial optical properties of light extraction layer easily with a simple masking and irradiation process, and it could maximize the outcoupling efficiency of OLEDs with proper manipulation. We believe that this solution-based PDLC film approach can play a critical role in future roll-to-roll processed flexible OLEDs.

Author Contributions: Conceptualization, Y.C. and S.L.; methodology, Y.C., J.-H.L., A.G. and S.L.; validation, Y.C. and S.L.; formal analysis, Y.C., J.-H.L., A.G. and S.L.; investigation, Y.C., J.-H.L., J.-S.G., H.-S.A., E.-J.J., S.-Y.B. and S.L.; resources, Y.C.; data curation, S.L.; writing—original draft preparation, S.L.; writing—review and editing, Y.C., J.-H.L., J.-S.G. and A.G.; visualization, S.L.; supervision, Y.C.; project administration, Y.C.; funding acquisition, Y.C. All authors have read and agreed to the published version of the manuscript.

Funding: This research was funded by the Basic Science Research Program through the NRF funded by the Ministry of Education (No. 2018R1A6A1A03026005) and This work was supported by "Regional Innovation Strategy (RIS)" through the National Research Foundation of Korea (NRF) funded by the MOE (2021RIS-004).

Institutional Review Board Statement: Not applicable.

Informed Consent Statement: Not applicable.

Data Availability Statement: Data are contained within the article.

Conflicts of Interest: The authors declare no conflict of interest.

References

1. Bauri, J.; Choudhary, R.B.; Mandal, G. Recent advances in efficient emissive materials-based OLED applications: A review. *J. Mater. Sci.* **2021**, *56*, 18837–18866. [CrossRef]
2. Sun, N.; Jiang, C.; Li, Q.; Tan, D.; Bi, S.; Song, J. Performance of OLED under mechanical strain: A review. *J. Mater. Sci. Mater. Electron.* **2020**, *31*, 20688–20729. [CrossRef]
3. Huang, Y.; Hsiang, E.L.; Deng, M.-Y.; Wu, S.-T. Mini-LED, Micro-LED and OLED displays: Present status and future perspectives. *Light Sci. Appl.* **2020**, *9*, 105. [CrossRef]
4. Huang, Y.; Tan, G.; Gou, F.; Li, M.-C.; Lee, S.-L.; Wu, S.-T. Prospects and challenges of mini-LED and micro-LED displays. *J. Soc. Inf. Disp.* **2019**, *27*, 387–401. [CrossRef]
5. Singh, R.; Unni, K.N.; Solanki, A. Improving the contrast ratio of OLED displays: An analysis of various techniques. *Opt. Mater.* **2012**, *34*, 716–723. [CrossRef]
6. Salehi, A.; Fu, X.; Shin, D.H.; So, F. Recent Advances in OLED Optical Design. *Adv. Funct. Mater.* **2019**, *29*, 1808803. [CrossRef]
7. Gather, M.C.; Reineke, S. Recent advances in light outcoupling from white organic light-emitting diodes. *J. Photon. Energy* **2015**, *5*, 057607. [CrossRef]
8. Mann, V.; Rastogi, V. Dielectric nanoparticles for the enhancement of OLED light extraction efficiency. *Opt. Commun.* **2017**, *387*, 202–207. [CrossRef]
9. Lee, J.-S.; Ko, J.-H.; Park, J.-H.; Lee, J.-W. Simulation Study on the Effect of the Emitter Orientation and Photonic Crystals on the Outcoupling Efficiency of Organic Light-Emitting Diodes. *J. Opt. Soc. Korea* **2014**, *18*, 732–738. [CrossRef]
10. Song, J.; Kim, K.-H.; Kim, E.; Moon, C.-K.; Kim, Y.-H.; Kim, J.-J.; Yoo, S. Lensfree OLEDs with over 50% external quantum efficiency via external scattering and horizontally oriented emitters. *Nat. Commun.* **2018**, *9*, 3207. [CrossRef] [PubMed]
11. Flämmich, M.; Michaelis, D.; Danz, N. Accessing OLED emitter properties by radiation pattern analyses. *Org. Electron.* **2011**, *12*, 83–91. [CrossRef]
12. Zhou, L.; Dong, X.-X.; Lv, G.-C.; Chen, J.; Shen, S. Fabrication of concave microlens array diffuser films with a soft transparent mold of UV-curable polymer. *Opt. Commun.* **2014**, *342*, 167–172. [CrossRef]

13. Han, J.-H.; Moon, J.; Cho, H.; Shin, J.-W.; Chu, H.Y.; Lee, J.-I.; Cho, N.S.; Lee, J. Luminescence enhancement of OLED lighting panels using a microlens array film. *J. Inf. Disp.* **2018**, *19*, 179–184. [CrossRef]
14. Gasonoo, A.; Lee, Y.-S.; Yoon, J.-H.; Sung, B.-S.; Choi, Y.; Lee, J.; Lee, J.-H. Outcoupling efficiency enhancement of a bot-tom-emitting OLED with a visible parylene film. *Opt. Express* **2020**, *20*, 26724–26732. [CrossRef] [PubMed]
15. Kim, A.; Huseynova, G.; Lee, L.; Lee, J.-H. Enhancement of out-coupling efficiency of flexible organic light-emitting diodes fabricated on an MLA-patterned parylene substrate. *Org. Electron.* **2019**, *71*, 246–250. [CrossRef]
16. Yeom, S.; Kim, H.; Kim, K.; Joo, C.W.; Cho, H.; Cho, H.; Choi, S.; Lee, W.J.; Jung, Y.S.; Kwon, B.-H.; et al. Surface wrinkle formation by liquid crystalline polymers for significant light extraction enhancement on quantum dot light-emitting diodes. *Opt. Express* **2020**, *28*, 26519–26530. [CrossRef] [PubMed]
17. Lee, J.H.; Kang, S.; Park, N.-M.; Shin, J.-W.; Joo, C.W.; Lee, J.; Ahn, S.-D.; Kang, S.-Y.; Moon, J. Porous cellulose paper as a light out coupling medium for organic light-emitting diodes. *J. Inf. Disp.* **2018**, *19*, 171–177.
18. Shin, C.-H.; Shin, E.Y.; Kim, M.-H.; Lee, J.-H.; Choi, Y. Nanoparticle scattering layer for improving light extraction efficiency of organic light emitting diodes. *Opt. Express* **2015**, *23*, A133–A139. [CrossRef]
19. Lee, K.; Shin, J.-W.; Park, J.-H.; Lee, J.; Joo, C.W.; Lee, J.-I.; Cho, D.-H.; Lim, J.T.; Oh, M.-C.; Ju, B.-K.; et al. A Light Scattering Layer for Internal Light Extraction of Organic Light-Emitting Diodes Based on Silver Nanowires. *ACS Appl. Mater. Interfaces* **2016**, *8*, 17409–17415. [CrossRef]
20. Jiang, J.; McGraw, G.; Ma, R.; Brown, J.; Yang, D.-K. Selective scattering polymer dispersed liquid crystal film for light enhancement of organic light emitting diode. *Opt. Express* **2017**, *25*, 3327–3335. [CrossRef]
21. Saeed, M.H.; Zhang, S.; Cao, Y.; Zhou, L.; Hu, J.; Muhammad, I.; Xiao, J.; Zhang, L.; Yang, H. Recent Advances in The Polymer Dispersed Liquid Crystal Composite and Its Applications. *Molecules* **2020**, *25*, 5510. [CrossRef] [PubMed]
22. Gasonoo, A.; Ahn, H.-S.; Lee, J.; Kim, M.-H.; Lee, J.-H.; Choi, Y. Polymer Dispersed Liquid Crystal for Enhanced Light Out-Coupling Efficiency of Organic Light Emitting Diodes. *J. IKEEE* **2020**, *24*, 140–146.
23. Saxena, K.; Jain, V.K.; Mehta, D.S. A review on the light extraction techniques in organic electroluminescent devices. *Opt. Mater.* **2009**, *32*, 221–233. [CrossRef]
24. Jiang, J.; McGraw, G.; Ma, R.; Brown, J.; Yang, D.-K. 48-4: Selective Scattering of PDLC film and Its Application in OLED. *SID Symp. Dig. Tech. Pap.* **2017**, *48*, 727–730. [CrossRef]
25. Yuan, W.; Li, L.H. Fabrication of Microlens Array and Its Application: A Review. *Chin. J. Mech. Eng.* **2018**, *31*, 16. [CrossRef]
26. Pane, S.; Caporusso, M.; Hakemi, H. Haze and opacity control in polymer dispersed liquid crystal (PDLC) films with phase separation method. *Liq. Cryst.* **1997**, *23*, 861–867. [CrossRef]
27. Seo, J.-H.; Han, N.-S.; Shim, H.-S.; Kwon, J.-H.; Song, J.-K. Phosphorescence Properties of Ir(ppy)3 Films. *Bull. Korean Chem. Soc.* **2011**, *32*, 1415–1418. [CrossRef]

molecules

Article

Preparation and Characterization of Bilayer Polymer-Dispersed Liquid Crystals Doped with Gd_2O_3 Nanoparticles and Rhodamine B Base Fluorescent Dye

Yongle Wu [1,2,†], Yuzhen Zhao [1,†], Xun Li [1], Hong Gao [3], Zhun Guo [1], Dong Wang [2,*], Yi Luan [2] and Lei Wang [4]

1 Xi'an Key Laboratory of Advanced Photo-Electronics Materials and Energy Conversion Device, School of Electronic Information, Xijing University, Xi'an 710123, China; wuyongle1999@163.com (Y.W.); zyz19870226@163.com (Y.Z.); lixun1999814@163.com (X.L.); guozhun@xijing.edu.cn (Z.G.)
2 Department of Materials Physics and Chemistry, School of Materials Science and Engineering, University of Science and Technology Beijing, Beijing 100083, China; yiluan@ustb.edu.cn
3 Division of Material Engineering, China Academy of Space Technology, Beijing 100094, China; gaohong_cast@sina.com
4 Key Laboratory of Chemical Additives for China National Light Industry, Shaanxi Key Laboratory of Chemical Additives for Industry, College of Chemistry & Chemical Engineering, Shaanxi University of Science & Technology, Xi'an 710021, China; wanglei@sust.edu.cn
* Correspondence: wangdong@ustb.edu.cn
† These authors contributed equally to this work.

Abstract: Using the polymerization-induced phase separation (PIPS) method, bilayer polymer-dispersed liquid crystal (PDLC) films with a PDLC-PVA-PDLC structure were prepared in this work. It was found that all PDLC performance indexes were affected by polymer mesh size after comparing the microscopic morphology and electro-optical properties of samples with different monomer ratios. Gd_2O_3 nanoparticles and rhodamine B base fluorescent dyes introduced into the bilayer PDLC optimized the samples' electro-optical properties and developed new functionalities. In addition, the bilayer PDLC doped with Gd_2O_3 and rhodamine B base held excellent progressive driving functions as well as stable durability properties. Samples doped with Gd_2O_3 nanoparticles and rhodamine B base also produced excellent anti-counterfeiting effects under UV irradiation at different angles, further exploiting the application potential of PDLC.

Keywords: polymer-dispersed liquid crystal; bilayer; electro-optical properties; Gd_2O_3 nanoparticles; fluorescent dyes

1. Introduction

As a typical liquid crystal-based optical film, polymer-dispersed liquid crystal (PDLC) consists of a polymer substrate with embedded micrometer- or nanometer-sized liquid crystal (LC) microdroplets [1]. PDLC films exhibit an opaque state under unelectrified conditions due to the random orientation of LC microdroplets [2]. In turn, when a sufficiently large voltage is applied, LC microdroplets align themselves along the direction of the electric field and cause PDLCs to reach a transparent state with high transmittance [3]. Based on the above features, PDLC has been widely used in smart windows [4–6], displays [7,8], and optical switches [9–11]. Furthermore, PDLC is favored in low energy-consuming buildings [12–14], semiconductors [15], field-effect transistors [16], and solar concentrators [17].

In previous studies, functionalization and performance optimization of PDLC were achieved by optimizing polymerization conditions, adjusting precursor ratios [18,19], and doping with additional [17] special properties. Deng et al. [20] designed and prepared an ultrafast, switchable, and passive radiative cooling smart window based on PDLC that exhibited enhanced infrared emissivity in the 8–13 μm wavelength range. The novel

PDLC/GO nanocomposite designed by Cheng et al. [21] responded to NIR-Vis-UV light and could be applied to soft actuators and optomechanical systems driven directly by sunlight. PDLCs accomplish fast switching speeds, large optical modulation, and robust mechanical stability using silver nanowire micromesh as electrodes, as demonstrated by Zhang et al. [22]. Shivaraja et al. [23] found that doping octadecylamine functionalized single-walled carbon nanotubes (ODA-SWCNTs) in PDLC could reduce threshold voltage and response time. The photoinitiator concentration directly affected the phase separation process, which determined the morphology and electro-optical properties of PDLCs, according to Nasir et al. [24]. In addition, other fluorescent dyes and inorganic nanoparticles were doped into PDLC for modification. Reticular nanofiber films containing the fluorescent material ST18 were blended into PDLC and improved electro-optical and fluorescent properties, according to Shi et al. [25] Electro-optical properties were enhanced by doping SiO_2 nanoparticles into PDLC by Kim et al. [26]. Katariya-Jain et al. found that carbon nanoparticle doping could enhance thermo-electro-optical and dielectric properties in PDLC [27].

To compensate for the shortcomings of monolayer materials, multilayer [28,29] structures have become a research hotspot with applications to a variety of fields in recent years. Using a layer-by-layer method, Khoi et al. [30] deposited a polymer multi-layer membrane consisting of poly allylamine hydrochloride and poly styrene sulfonate on the cation-exchange membrane, which achieved selective recovery of Na ions. Multilayer films prepared by Culebras et al. [31] had high conductivity and Seebeck coefficients with alternately depositing layers of aqueous solutions of PEDOT: NPs and double-walled carbon nanotubes (DWNT) stabilized with PEDOT: PSS via a layer-by-layer methodology. The multilayer composite electrolyte designed and prepared by Park et al. [32] displayed excellent cyclic stability for application in lithium–sulfur batteries. The multilayer structure consisting of paper, TPS and PLA, PBAT, PHBV, or PVOH by Eslami et al. [33] not only improved the dry and wet strength of the paper but also showed considerable resistance to oil and grease. Low dielectric properties of monolayer in-plane heterostructures prevent the generation of high concentrations of thermally excited carriers from doped impurities. Ogura et al. [34] solved this problem by developing multilayer in-plane heterostructures.

However, conventional monolayer PDLC suffers from certain drawbacks, such as difficulty accessing the intermediate state between the on-state and off-state. It greatly limits the further development of PDLC. In this study, bilayer PDLC films were prepared by introducing the concept of multilayer structures into PDLC, which can access the intermediate state and achieve progressive driving functions. In addition, Gd_2O_3 nanoparticles and fluorescent dyes were doped into the bilayer PDLC to further improve its electro-optical properties and confer functionalization. The stable PDLC–PVA–PDLC structure also enables the bilayer PDLC to perform well in durability tests, and a longer service life can meet liquid crystal optical film application requirements.

2. Experimental

2.1. Materials

SLC1717 (n_o = 1.519, n_e = 1.720, Δn = 0.201, T_c = 92 °C), the nematic phase liquid crystal used in this experiment, was purchased from Shijiazhuang Chengzhi Yonghua Display Materials Co. (Shijiazhuang, China). As a mixture of UV polymerizable monomers, UV6301 was bought from Kuer Industries (Shanghai) Co. (Shanghai, China). As a crosslinker and fluorescent dye, 1,4 Butanediol dimethacrylate (BDDMA) and Rhodamine B base were purchased from Beijing Bailingwei Technology Co. (Beijing, China), respectively. The free radical photoinitiator 2,2-dimethopxy-2-phenylacetophenone (IRG651) was purchased from Anhui Zesheng Technology Co. (Anqing, China).

Gd_2O_3 nanoparticles (diameter 40–60 nm), the inorganic nanoparticles used in this experiment, were acquired from Alab (Shanghai, China) Chemical Technology Co. (Shanghai, China). Oleic acid, a modifier for the nanoparticles, was bought from Sinopharm Chemical Reagent Co. (Shanghai, China). The specific chemical structures of materials used in this study are presented in Figure 1. The Gd_2O_3 nanoparticles were surface-modified using oleic acid to inhibit their agglomeration and uniformly disperse them in the LC/monomer mixtures.

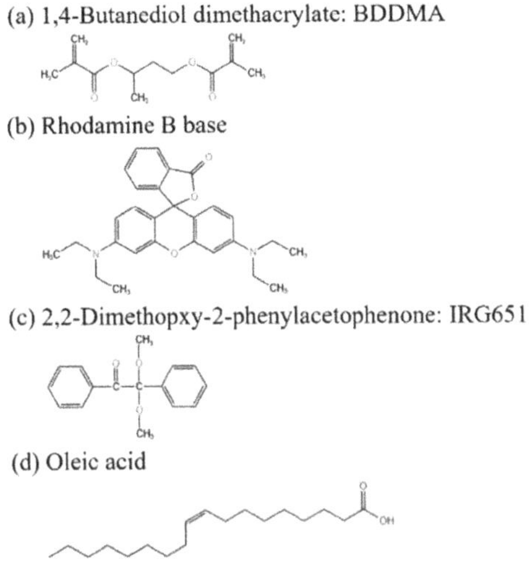

Figure 1. The specific chemical structure of the materials used.

Details concerning the surface modification treatment of Gd_2O_3 nanoparticles are described as follows. The Gd_2O_3 nanoparticles (10 mmol) were scattered in tetrahydrofuran (20 mL) and mechanically stirred (300–600 rpm) at 65 °C for 30 min for homogeneous mixing. The oleic acid (5 mL)/tetrahydrofuran (10 mL) mixture was poured into the Gd_2O_3 solution and continuously stirred for 30 min under heating conditions. Oleic acid (2 mL) was mixed into the Gd_2O_3 solution for 5 min and the procedure was repeated five times. For adequate modification, the above mixture was mechanically stirred (300–600 rpm) at 85 °C for 2 h. The heating process described above must be conducted under condensing conditions to avoid tetrahydrofuran volatilization. Afterward, the heat source was removed to cool the mixture to room temperature. The precipitate was obtained after centrifuging (5000 rpm, 5 min), separating the above blend, and washing five times with anhydrous ethanol for primary modified Gd_2O_3 nanoparticles. The above primary modified Gd_2O_3 nanoparticles were solubilized in the oleic acid (5 mL)/hexane (20 mL) mixture and mechanically stirred (800~1000 rpm) for 2 h. The precipitate was obtained after centrifuging (5000 rpm, 10 min), separating the above blend, washing five times with anhydrous ethanol, and then drying to obtain secondary modified Gd_2O_3 nanoparticles.

2.2. Sample Preparation

PVA films are ideal for preparing bilayer PDLC interlayers due to their excellent mechanical properties. In addition, PVA film is immiscible with LC, which is critical for preparing bilayer PDLC. The procedures for specific preparation are as follows.

1. Preparing precursors: LC/monomer/initiator blends in different proportions were shaken for 3 min, stirred for 5 min, and sonicated for 15 min to produce a uniform phase. Specific percentages are described in Tables 1–3;
2. Preparing monolayer PDLC: LC cells were prepared using an 8 μm thick polyimide film as a spacer to hold two single-sided conductive glass substrates apart. Mixtures a1~a6 in Table 1 were injected into LC cells by capillary action, and monolayer PDLC samples were obtained after polymerization. The polymerization time, light intensity, and temperature for this set of samples were 6 min, 4.5 mW/cm^2, and 35 °C, respectively;

3. Preparing PVA films: A layer of aqueous PVA solution (0.05 wt%) was placed on the glass substrate with surface treatment using surface tension. Glass substrates with PVA films adhered to the surface were obtained by storing them at a temperature of 75 °C for 3 h;
4. Preparing bilayer PDLC: The ratios of various substances in the upper and lower layers of the bilayer PDLC are shown in Tables 2 and 3, respectively. An 8 μm thick polyimide film was used to separate the glass substrate with PVA film on the surface from the unilateral conductive glass substrate and fixed to obtain the LC cell. The upper PDLC precursor was injected into the LC cell by capillary action, and PDLC-PVA composite films were obtained after curing. The glass substrates with PDLC-PVA composite film on the surface were obtained by peeling off the surface-treated glass substrates. The glass substrates attached with PDLC-PVA composite film on the surface were separated from the one-sided conductive glass substrates using 8 μm thick polyimide films to obtain new LC cells. The lower PDLC precursors were injected into new LC cells by capillary action, and bilayer PDLC films with a PDLC–PVA–PDLC structure were acquired after polymerization.

Table 1. The composition of Group A [a].

Sample	Weight Percentage (wt%)		
	SLC1717	UV6301	BDDMA
Group A			
a1	50	50	0
a2	50	49	1
a3	50	48	2
a4	50	47	3
a5	50	46	4
a6	50	45	5

[a] The weight of the photo-initiator IRG651 is 0.5% of the total weight.

Table 2. The upper compositions of Groups B and C.

Sample	Weight Percentage (wt%)				
	SLC1717	UV6301	BDDMA	Gd_2O_3 Nanoparticles	Rhodamine B Base
Group B					
b1	50	49	1	0.0	0.0
b2	50	49	1	0.2	0.0
b3	50	49	1	0.4	0.0
b4	50	49	1	0.6	0.0
b4	50	49	1	0.8	0.0
b6	50	49	1	1.0	0.0
Group C					
c1	50	49	1	0.2	0.0
c2	50	49	1	0.2	0.2
c3	50	49	1	0.2	0.4
c4	50	49	1	0.2	0.6
c5	50	49	1	0.2	0.8
c6	50	49	1	0.2	1.0

Among them, the variables of Groups B and C were the Gd_2O_3 nanoparticle and Rhodamine B base content, respectively, as shown in Tables 2 and 3.

Table 3. The lower composition of Groups B and C.

Sample	Weight Percentage (wt%)				
	SLC1717	UV6301	BDDMA	Gd_2O_3 Nanoparticles	Rhodamine B Base
Group B					
b1	50	47	3	0.0	0.0
b2	50	47	3	0.2	0.0
b3	50	47	3	0.4	0.0
b4	50	47	3	0.6	0.0
b4	50	47	3	0.8	0.0
b6	50	47	3	1.0	0.0
Group C					
c1	50	47	3	0.2	0.0
c2	50	47	3	0.2	0.2
c3	50	47	3	0.2	0.4
c4	50	47	3	0.2	0.6
c5	50	47	3	0.2	0.8
c6	50	47	3	0.2	1.0

2.3. Characterization

To verify whether oleic acid was attached to the surface of Gd_2O_3 nanoparticles, its chemical composition was characterized using Fourier-Transform Infrared Spectrometry (FTIR, INVENIO S, Bruker Optics GmbH & Co. KG, Ettlingen, Germany).

Electro-optical characteristics are one of the most critical metrics and can usually measure the application potential of PDLC. In this experiment, the electro-optical curve, response time curve, and contrast ratio were evaluated with the Liquid Crystal Comprehensive Parameter Tester (LCT-5016C, Beijing LCD Engineering Research and Development Center, Beijing, China). Additionally, other vital variables such as threshold voltage (V_{th}), saturation voltage (V_{sat}), off-state transmittance (T_{off}), and off-state response time (t_{off}) were also obtained from electro-optical characterization tests. Herein, V_{th} and V_{sat} were attained once the transmittance of the bilayer PDLC reached 10% and 90% of its maximum value, respectively. The ratio of on-state to off-state transmittance is called the contrast ratio (CR).

The polymer matrix morphology was observed by scanning electron microscopy (SEM, ZEISS SUPRA55, Carl Zeiss AG, Oberkochen, Germany). Samples were pretreated prior to observation as follows: PDLC samples were immersed in cyclohexane for 15 d at room temperature to clear all LC molecules. The cyclohexane was renewed every three days during the soaking process to ensure the adequate deletion of LC molecules. After soaking, the samples were dried at 60 °C for 1 h to remove residual cyclohexane from the surface. After drying, the samples were sprayed with gold under a vacuum for SEM visualization.

The fluorescence emission spectra of bilayer PDLC samples doped with Rhodamine B base were characterized using a steady-state transient fluorescence spectrometer (Edinburgh Instruments, Livingston, UK, FLS 1000), and fluorescence emission curves were obtained for different doping concentrations.

The samples were tested for transmittance at different wavelengths with the Lambda 950 UV/VIS/NIR spectrophotometer (Perkin–Elmer, Waltham, MA, USA) at room temperature and transmittance–wavelength curves were obtained for the bilayer PDLC at different voltages.

3. Results and Discussions

3.1. Effect of the Crosslinker Content on the Property of Monolayer PDLC

Crosslinking agents connect polymer chains and reduce the size of the polymer mesh. The SEM image of the polymer mesh with variations in crosslinker BDDMA content is shown in Figure 2. As the crosslinker content increased, the number of crosslinking points in the polymer network also increased, leading to a decrease in the size of the polymer mesh.

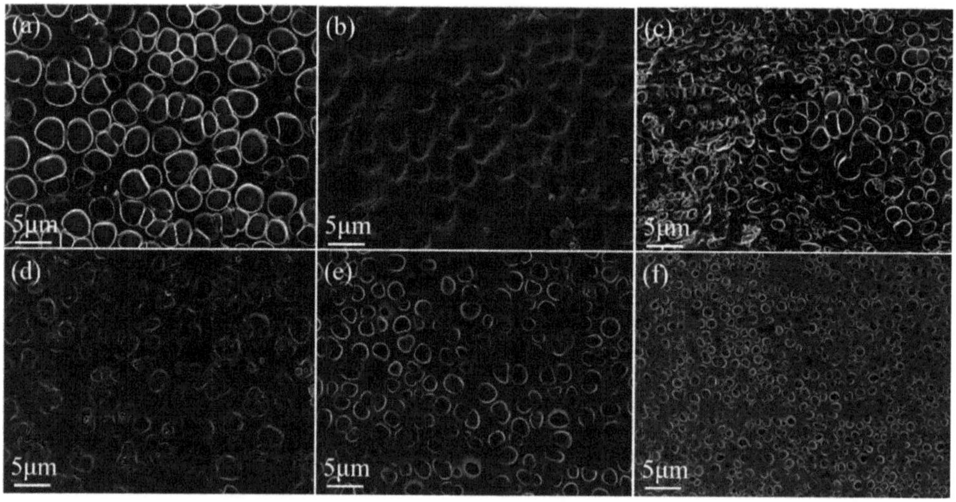

Figure 2. SEM images of PDLC cells. (**a**) 0 wt% BDDMA; (**b**) 1 wt% BDDMA; (**c**) 2 wt% BDDMA; (**d**) 3 wt% BDDMA; (**e**) 4 wt% BDDMA; (**f**) 5 wt% BDDMA. 3000× magnification.

Variations in polymer micromorphology led to changes in the electro-optical properties of PDLC, as depicted in Figure 3. The anchoring force on LC droplets increases with smaller polymer meshes, which shifts the transmittance-voltage curve to the right and raises V_{th} and V_{sat}, as demonstrated in Figure 3a,b. V_{th} and V_{sat} were enlarged from 5.3 V and 10.0 V to 26.4 V and 66.9 V, respectively, when BDDMA content increased from 0 to 5 wt%. The growth of V_{sat} was greater than V_{th}, increasing ΔV from 4.7 V to 40.5 V. Changes in polymer morphology can cause variation in the refractive index match between the polymer and LC microdroplets, altering the T_{off} and CR of the PDLC, as illustrated in Figure 3c. The refractive index match between the polymer and the LC microdroplets was weakened with increasing BDDMA content, causing a gradual decrease in T_{off} from 13.6% to 4.2%. CR gradually grew from 7.3 to 23.8 as the CR tendency became contrary to T_{off}. The larger the anchoring force, the simpler it is for the LC microdroplets to recover from an ordered state to a disordered state. The t_{off} decreased from 78.8 ms to 7.1 ms as depicted in Figure 3d.

3.2. Gd_2O_3 Nanoparticles Doped Bilayer PDLC

3.2.1. Modification of Gd_2O_3 Nanoparticles

Gd_2O_3 nanoparticles were modified with oleic acid to prevent agglomeration and promote uniform dispersion in the precursors of LC/UV polymerizable monomers. The FTIR spectra of Gd_2O_3 nanoparticles before and after modification are shown in Figure 4. Compared to pre-modification, the modified Gd_2O_3 nanoparticles showed a characteristic peak at 2900 cm^{-1}, which represents the stretching vibration peak of the methylene group. It can be seen that oleic acid was successfully modified on the surface of Gd_2O_3 nanoparticles.

3.2.2. Effect of Gd_2O_3 Nanoparticle Content on the Properties of Bilayer PDLC

Gd_2O_3 nanoparticles affect polymers' microscopic morphology, as indicated in Figures 5 and 6. Gd_2O_3 nanoparticles could access LC microdroplets and increase their size, thus changing the polymer's micromorphology. As Gd_2O_3 nanoparticle content increased, the size of LC microdroplets also increased, resulting in a gradual increase in polymer mesh size.

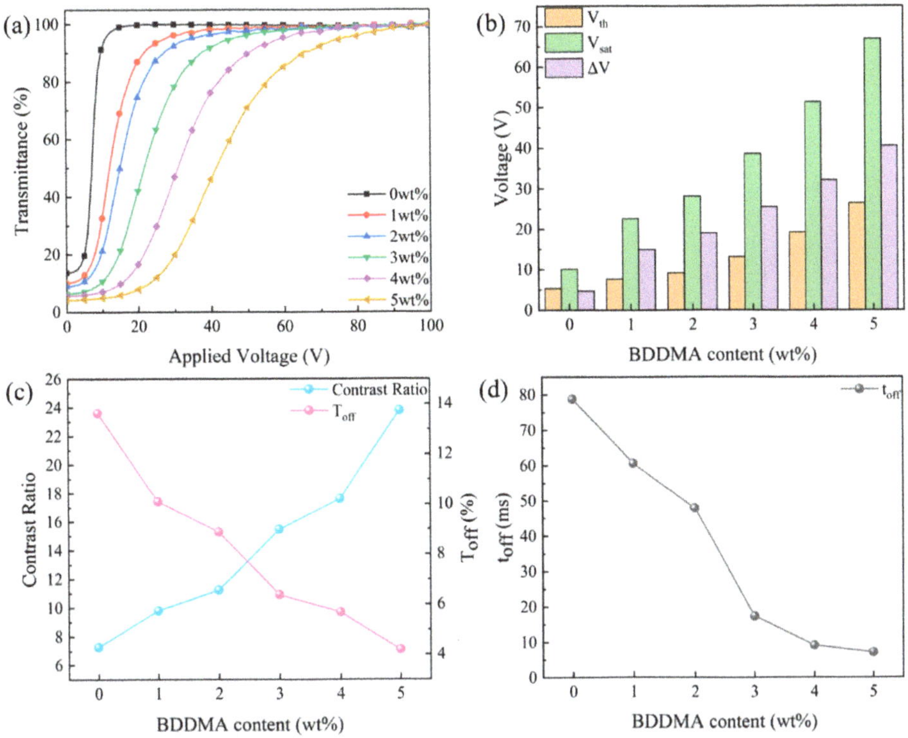

Figure 3. The effect of cross-linker (BDDMA) content on electro-optical properties: (**a**) voltage-transmittance curve; (**b**) threshold voltage (V_{th}), saturation voltage (V_{sat}), and ΔV; (**c**) contrast ratio (CR) and off-state transmittance (T_{off}); (**d**) response time (t_{off}).

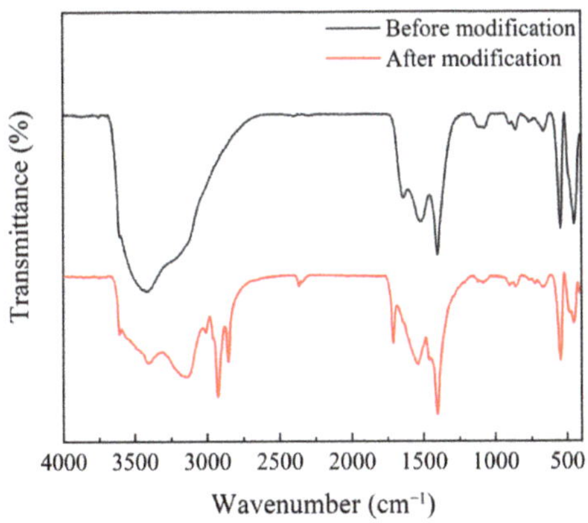

Figure 4. Fourier-transform infrared (FTIR) spectra of Gd_2O_3 nanoparticles before and after modification.

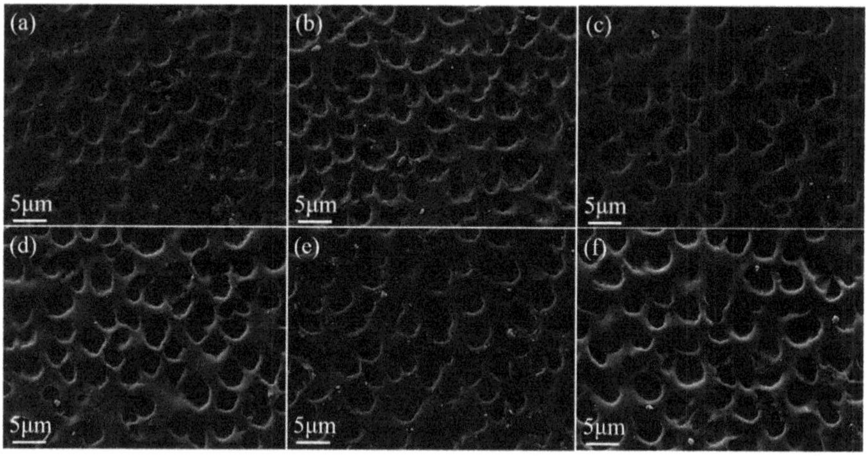

Figure 5. SEM images of upper-layer PDLC cells. (**a**) 0.0 wt%; (**b**) 0.2 wt%; (**c**) 0.4 wt%; (**d**) 0.6 wt%; (**e**) 0.8 wt%; (**f**) 1.0 wt% Gd_2O_3 nanoparticles doped. 3000× magnification.

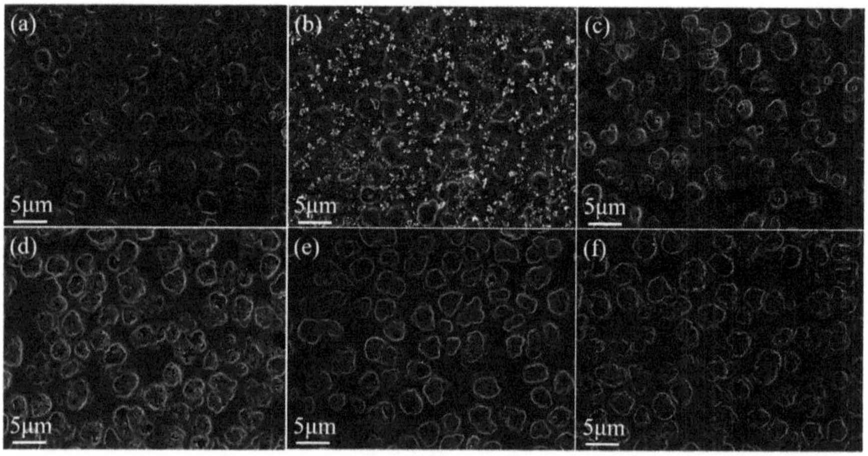

Figure 6. SEM images of lower-layer PDLC cells. (**a**) 0.0 wt%; (**b**) 0.2 wt%; (**c**) 0.4 wt%; (**d**) 0.6 wt%; (**e**) 0.8 wt%; (**f**) 1.0 wt% Gd_2O_3 nanoparticles doped; 3000× magnification.

As shown in Figure 7, the fraction of Gd_2O_3 nanoparticles significantly impacts the electro-optical properties of the bilayer PDLC. Electro-optical curves migrated to the right and then left as the size of the polymer mesh increased in Figure 7a. While the percentage of Gd_2O_3 nanoparticles enlarged from 0 to 0.4 wt%, the anchoring force applied to the LC droplets progressively declined, bringing about reductions in V_{th} and V_{sat} of the bilayer PDLC from 30.2 V and 79.8 V to 17.8 V and 50.4 V, respectively. Since the reduced value of V_{sat} was greater than V_{th}, ΔV decreased, according to Figure 7b. With the increase in Gd_2O_3 nanoparticles to 1 wt%, V_{th} and V_{sat} gradually increased to 36.2 V and 83.9 V, while ΔV was roughly kept at 50 V.

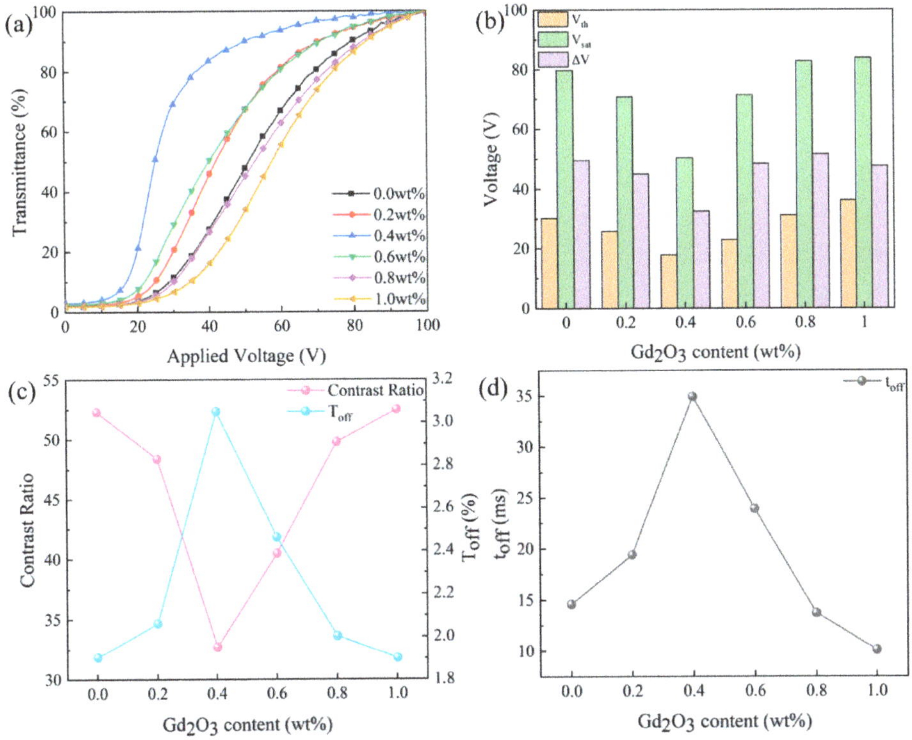

Figure 7. The effect of Gd_2O_3 nanoparticle content on electro-optical properties. (**a**) voltage-transmittance curve; (**b**) threshold voltage (V_{th}), saturation voltage (V_{sat}), and ΔV; (**c**) contrast ratio (CR) and off-state transmittance (T_{off}); (**d**) response time (t_{off}).

Additionally, the fraction of Gd_2O_3 nanoparticles impacted the T_{off} and CR of bilayer PDLC, as demonstrated in Figure 7c. The increase in Gd_2O_3 nanoparticle content from 0 to 0.4 wt% improved the T_{off} from 1.9% to 3.0%. However, a further increase in Gd_2O_3 nanoparticle content to 1.0 wt% brought about a decrease in T_{off} to 1.9%. Given that the CR's tendency was the opposite of T_{off}, the CR first decreased from 52.3 to 32.71 and then increased to 52.5. Variations in Gd_2O_3 nanoparticle content affected the time required for LC droplets to transition from an ordered to a disordered state, as observed in Figure 7d. By increasing the Gd_2O_3 nanoparticle content from 0 to 0.4 wt%, the t_{off} increased from 14.6 ms to 34.9 ms. However, as the nanoparticle content was further extended to 1.0 wt%, the t_{off} decreased to 10.1 ms.

The EDS map of the sample with the optimal Gd_2O_3 nanoparticle content (b2) is shown in Figure 8. According to the Gd elemental distribution, the Gd_2O_3 nanoparticles are more uniformly distributed. In addition, the agglomeration of Gd_2O_3 nanoparticles is not obvious, which is a prerequisite for maintaining PDLC's excellent performance.

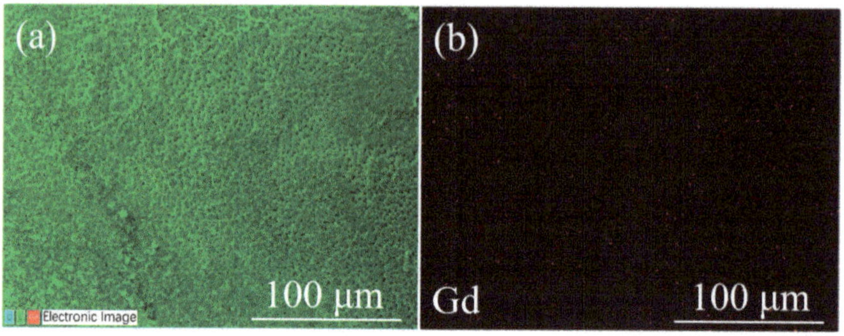

Figure 8. EDS mapping of the optimal sample (b2). (**a**) The distribution of all elements and (**b**) the distribution of Gd elements.

3.2.3. Progressive Driving Test of Bilayer PDLC Doped with Gd_2O_3 Nanoparticles

The change in transmittance was observed by applying different voltages to the sample and verifying the progressive driving performance of the bilayer PDLC. As the applied voltage increased, there was a stepwise growth in the sample transmittance, as shown in the physical diagram in Figure 9. The reason for this result is that the two monolayer PDLC samples have different drive voltages so both cannot be driven simultaneously. The cross-sectional SEM image of the bilayer PDLC is plotted in Figure 10, where one layer of the PDLC is easier to drive and the other is more difficult to drive.

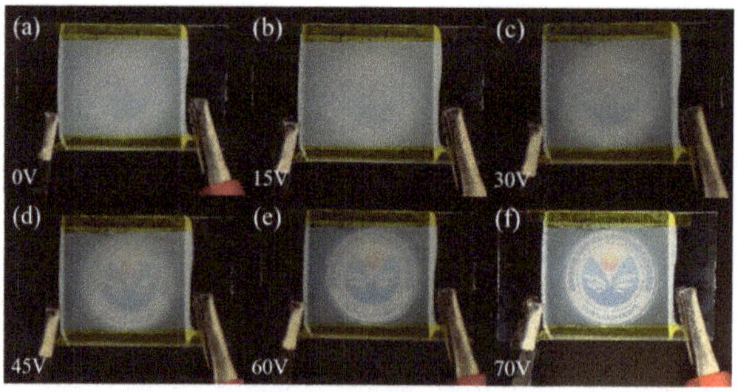

Figure 9. Physical plots of transmittance variation for the optimum sample at different voltages: (**a**) 0 V, (**b**) 15 V, (**c**) 30 V, (**d**) 45 V, (**e**) 60 V, and (**f**) 70 V, respectively.

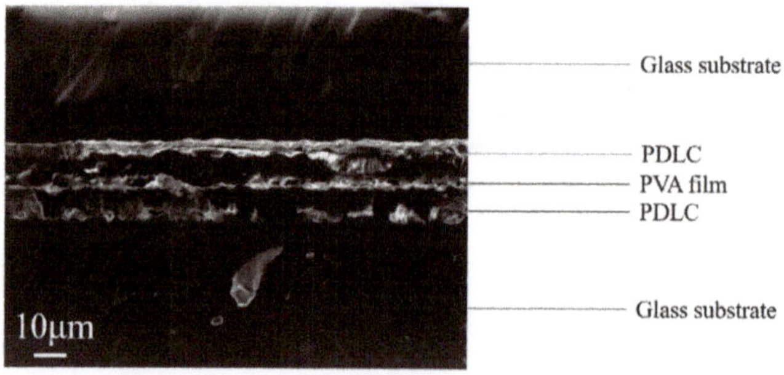

Figure 10. SEM image of a bilayer PDLC cross-section with a PDLC–PVA–PDLC structure.

3.3. Fluorescent Dye Rhodamine B Base Doped Bilayer PDLC

3.3.1. Effect of the Fluorescent Dye Rhodamine B Base on Bilayer PDLC Properties

In the set of experiments, different rhodamine B base levels were incorporated into the precursors to explore the influence of different rhodamine B base levels on bilayer PDLC performance. The impact of rhodamine B base contention on the electro-optical properties of bilayer PDLC is summarized in Figure 11. Fluorescent dyes can produce fluorescent effects and absorb UV light under UV irradiation, inducing a diminution of light radiation intensity in polymerized monomers. Consequently, the polymerization rate reduces and leads to an expansion of the polymer mesh and LC droplet sizes.

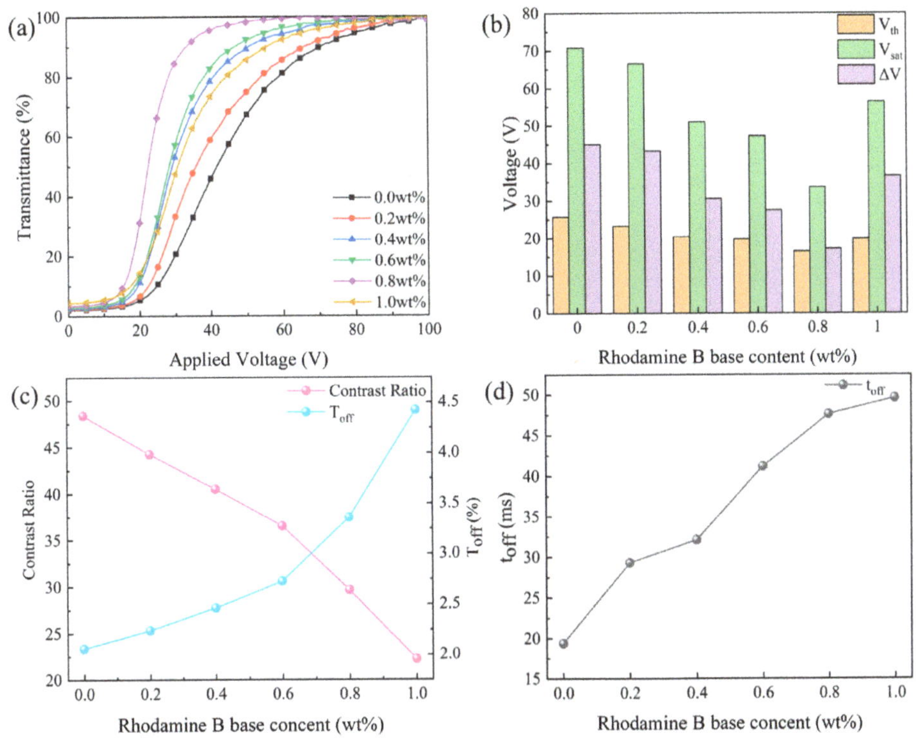

Figure 11. The effect of Rhodamine B base content on electro-optical properties. (**a**) Voltage-transmittance curve; (**b**) threshold voltage (V_{th}), saturation voltage (V_{sat}), and ΔV; (**c**) contrast ratio (CR) and off-state transmittance (T_{off}); (**d**) response time (t_{off}).

After the addition of Rhodamine B base proportion from 0 to 0.8 wt%, the transmittance-voltage curve moved to the left and caused a gradual reduction in V_{th} and V_{sat} from 25.8 V and 70.8 V to 16.5 V and 33.6 V. With the further increase in rhodamine B base content to 1.0 wt%, the transmittance–voltage curves moved to the right and V_{th} and V_{sat} increased to 19.8 V and 56.3 V, as illustrated in Figure 11a,b. The reason for this result is that as the fluorescent dye content increased, the anchoring force on the LC droplets decreased and became easier to drive. However, too much fluorescent dye content led to poor conductivity in the sample, which caused the driving voltage to increase.

Rhodamine B base content also affected the CR of bilayer PDLC, as depicted in Figure 11c. Since T_{off} gradually decreased with increasing rhodamine B base content, the CR gradually increased. The fluorescent dye increased the refractive index match between the polymer network and LC microdroplets, which increased the T_{off} of the bilayer PDLC and decreased the CR. In addition, rhodamine B base content affected the response time of bilayer PDLC, according to Figure 11d. As the anchoring force on LC microdroplets gradually decreased, the time required for their transition from the ordered to the disordered state increased, leading to an increase in t_{off}.

The fluorescence emission spectra and visual pictures of bilayer PDLC samples doped with different rhodamine B base percentages are depicted in Figures 12 and 13, respectively. There was a marked upward trend in the fluorescence emission intensity of bilayer PDLC samples as the rhodamine B base content increased. Additionally, there was a series of emission peaks within the metering range, positioned at 590 nm. Compared to no UV irradiation, the samples' orange coloration becomes more pronounced with the increase in rhodamine B base levels, as shown in the physical diagram in Figure 13. One explanation for this is that adding rhodamine B base to PDLC caused a strong fluorescence intensity, which was enhanced by increased fluorescent dye content.

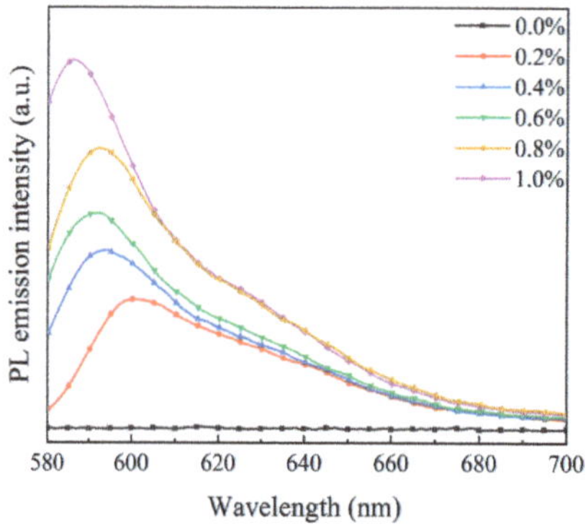

Figure 12. Fluorescence emission spectra of bilayer PDLC doped with rhodamine B base.

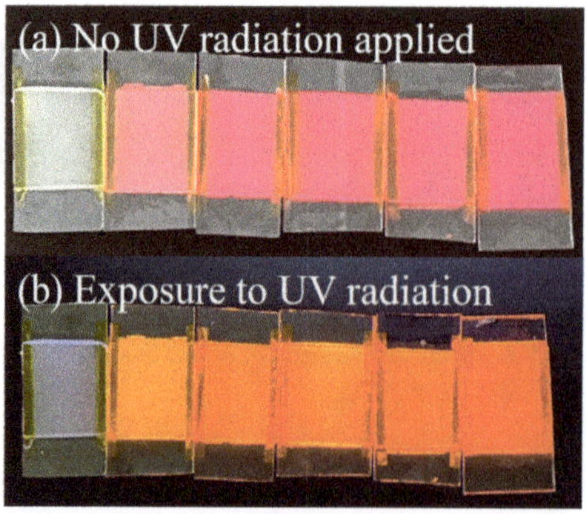

Figure 13. Physical image of bilayer PDLC doped with rhodamine B base. (**a**) No UV radiation applied, (**b**) Exposure to UV radiation.

3.3.2. Progressive Driving Test of Bilayer PDLC Doped with Rhodamine B Base

The physical plots of the transmittance variation of bilayer PDLC doped with rhodamine B base under UV irradiation at different voltages are demonstrated in Figure 14. With the increase of voltage, the transmittance of the bilayer PDLC increases gradually and the pattern at the bottom is clearer, which indicates that the bilayer PDLC sample has favorable progressive driving performance. In addition, due to the fluorescence effect of rhodamine B base, the color of the pattern through the sample changed accordingly.

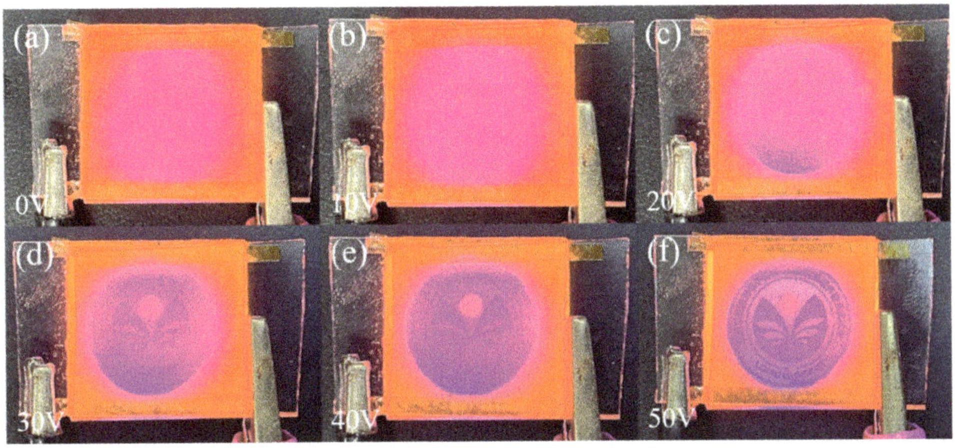

Figure 14. Physical plots of transmittance variation of the optimum sample doped with Rhodamine B base at different voltages: (**a**) 0 V, (**b**) 10 V, (**c**) 20 V, (**d**) 30 V, (**e**) 40 V, and (**f**) 50 V, respectively.

3.4. High Temperature and Strong UV Light Radiation Tests

The prepared bilayer PDLC samples b2 and c3 were tested for aging resistance in a long-term high temperature (50 °C) and strong UV (830 mW/cm^2) environment. The test samples were subjected to multiple tests at high temperatures and under UV light to test their electro-optical properties. Specific changes in V_{sat} and t_{off} are shown in Figure 15, which both increase with aging time. The rate of change in samples doped with rhodamine B base was less than the undoped ones, suggesting that doping with rhodamine B base helps PDLC performance under extreme conditions and has a better lifetime. The presence of rhodamine B base absorbs UV light and converts it to orange light, thereby reducing the damage caused by UV light to the polymer network and providing endless possibilities for future developments.

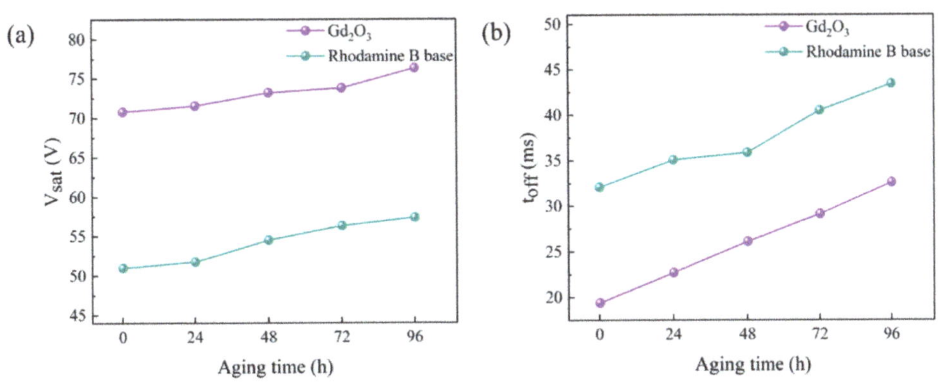

Figure 15. Effects of high temperature and strong ultraviolet radiation on (**a**) saturation voltage (V_{sat}) and (**b**) response time (t_{off}).

3.5. Application of Bilayer PDLC Doped with Rhodamine B Base in Multi-Angle Anti-Counterfeiting

Bilayer PDLC samples doped with rhodamine B base showed orange coloration by fluorescence emission under UV light irradiation. Setting a specific color at a specific location in the pattern can reveal another color when superimposed with the orange color emitted by the fluorescence of Rhodamine B. Using this feature, PDLC samples doped with rhodamine B base can be applied to the field of anti-counterfeiting. When voltage is applied, the pattern underneath the sample will show a special light after changing the angle of UV irradiation, as shown in Figure 16. When undoped with Rhodamine B, the pattern color is displayed normally. However, when doped with Rhodamine B, special colors develop at specific locations in the pattern. Using UV light to illuminate the sample at different angles will also have different color effects.

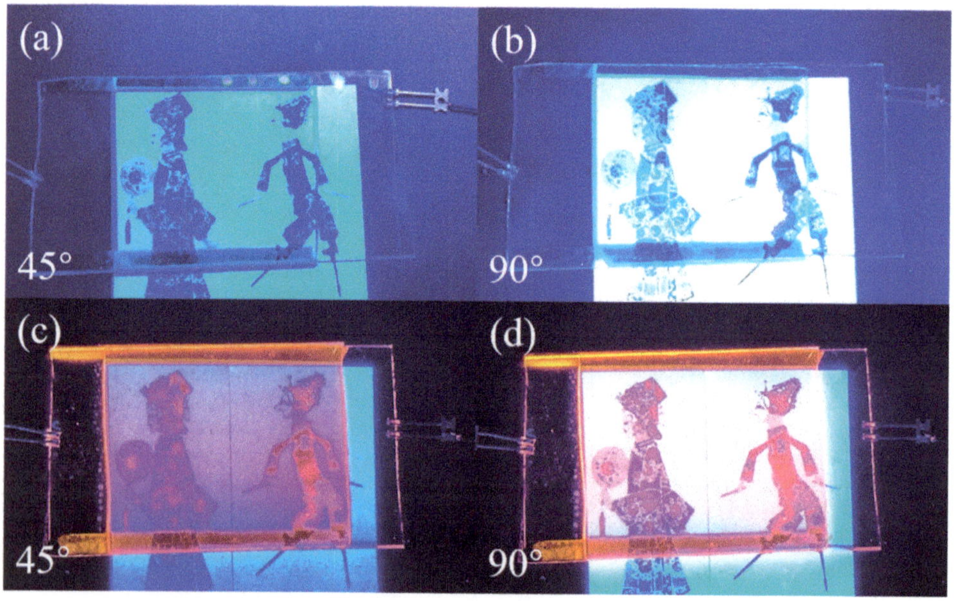

Figure 16. Application of bilayer PDLC in anti-counterfeiting. (**a**) Undoped rhodamine B base and UV 45° oblique light; (**b**) undoped rhodamine B base and UV 90° direct light; (**c**) doped rhodamine B base and UV 45° oblique light; (**d**) doped rhodamine B base and UV 90° direct light.

4. Conclusions

In conclusion, many problems in conventional PDLC devices, such as difficulty accessing the intermediate state between the on-state and the off-state under low-voltage conditions, can be solved by introducing Gd_2O_3 nanoparticles and rhodamine B base. According to our experimental results, the polymer network was affected by the ratio of LC/polymerizable monomer, which led to changes in electro-optical properties. In addition, the binding of Gd_2O_3 to LC microdroplets and the fluorescence emission from rhodamine B base also affected the polymer mesh, thus conferring excellent electro-optical properties on bilayer PDLCs. Compared to the original samples, samples obtained with Gd_2O_3 and rhodamine B base at the optimal content had superior electro-optical properties, with V_{th} and V_{sat} reduced by about 30% and 35%, respectively. The multi-angle anti-counterfeiting function can also be practiced by the doping of rhodamine B base, leading to innovative application areas for PDLC. Doping with Gd_2O_3 nanoparticles and fluorescent dye rhodamine B base not only improved electro-optical properties but also the functionalization of PDLC films. This study provides promising new ideas for preparing functional PDLC devices.

Author Contributions: Conceptualization, Y.W., X.L., D.W. and L.W.; data curation, Y.W.; formal analysis, Y.W. and Z.G.; funding acquisition, Y.Z.; investigation, Y.W. and H.G.; methodology, Y.W. and Y.L.; project administration, Y.Z.; resources, Y.Z., Z.G., D.W. and L.W.; software, Y.W., X.L. and H.G.; validation, Y.W. and X.L.; visualization, Y.W.; writing—original draft, Y.W.; writing—review and editing, Y.W., D.W. and Y.L. All authors have read and agreed to the published version of the manuscript.

Funding: This work was supported by the Natural Science Basic Research Plan in Shaanxi Province of China (Program No. 2024JC-YBMS-342), the Natural Science Foundation of Shaanxi Provincial Department of Education (Program No. 22JP100), Science and technology plan project of Xi'an (No. 22GXFW0150), National Natural Science Foundation of China (No. 51971182), Xi'an Science and Technology Plan Project of Shaanxi Province (No. 23LLRH0008), and the Youth Innovation Team of Shaanxi Universities.

Institutional Review Board Statement: Not applicable.

Informed Consent Statement: Not applicable.

Data Availability Statement: The data presented in this study are available in article.

Conflicts of Interest: The authors declare no conflict of interest.

References

1. Zhang, H.M.; Miao, Z.C.; Shen, W.B. Development of polymer-dispersed liquid crystals: From mode innovation to applications. *Compos. Part A Appl. Sci. Manuf.* **2022**, *163*, 107234. [CrossRef]
2. Hemaida, A.; Ghosh, A.; Sundaram, S.; Mallick, T.K. Simulation study for a switchable adaptive polymer dispersed liquid crystal smart window for two climate zones (Riyadh and London). *Energ. Build.* **2021**, *251*, 111381. [CrossRef]
3. Li, K.; Wang, J.W.; Cai, W.F.; He, H.L.; Cen, M.J.; Liu, J.X.; Luo, D.; Mu, Q.Q.; Gérard, D.; Liu, Y.J. Electrically switchable, polarization-sensitive encryption based on aluminum nanoaperture arrays integrated with polymer-dispersed liquid crystals. *Nano Lett.* **2021**, *21*, 7183–7190. [CrossRef] [PubMed]
4. Kumar, S.; Kang, D.; Nguyen, V.H.; Nasir, N.; Hong, H.; Kim, M.; Nguyen, D.C.; Lee, Y.J.; Lee, N.; Seo, Y. Application of titanium-carbide MXene-based transparent conducting electrodes in flexible smart windows. *ACS Appl. Mater. Interfaces* **2021**, *13*, 40976–40985. [CrossRef] [PubMed]
5. Kamal, W.; Li, M.M.; Lin, J.D.; Parry, E.; Jin, Y.H.; Elston, S.J.; Castrejón-Pita, A.A.; Morris, S.M. Spatially patterned polymer dispersed liquid crystals for image-integrated smart windows. *Adv. Opt. Mater.* **2022**, *10*, 2101748. [CrossRef]
6. Iluyemi, D.C.; Nundy, S.; Shaik, S.; Tahir, A.; Ghosh, A. Building energy analysis using EC and PDLC based smart switchable window in Oman. *Sol. Energy* **2022**, *237*, 301–312. [CrossRef]
7. Gong, S.; Cao, Y.; Fang, X.; Wang, Y.F.; Liu, Q.Z.; Gui, H.; Shen, C.F.; Cao, X.; Kim, E.S.; Zhou, C.W. Carbon nanotube macroelectronics for active matrix polymer-dispersed liquid crystal displays. *ACS Nano* **2016**, *10*, 10068–10074.
8. Manabe, H.; Date, M.; Takada, H.; Inamura, H. Techniques to reduce driving energy of 1-pixel displays. *IEEE Trans. Ind. Appl.* **2016**, *52*, 2638–2647. [CrossRef]
9. Zhang, C.; Wang, H.L.; Guan, S.; Guo, Z.H.; Zheng, X.X.; Fan, Y.J.; Wang, Y.; Qu, T.; Zhao, Y.B.; Chen, A.H.; et al. Self-powered optical switch based on triboelectrification-triggered liquid crystal alignment for wireless sensing. *Adv. Funct. Mater.* **2019**, *29*, 1808633. [CrossRef]
10. Van, M.C.W.; Janssen, R.H.C.; Broer, D.J.; Wilderbeek, H.T.A.; Bastiaansen, C.W.M. Polymer-filled nematics: A new class of light-scattering materials for electro-optical switches. *Adv. Mater.* **2000**, *12*, 753–757.
11. Baars, M.W.P.L.; Van, M.C.W.; Bastiaansen, C.W.M.; Broer, D.J.; Söntjens, S.H.M.; Meijer, E.W. A scattering electro-optical switch based on dendrimers dispersed in liquid crystals. *Adv. Mater.* **2000**, *12*, 715–719. [CrossRef]
12. Ghosh, A. Diffuse transmission dominant smart and advanced windows for less energy-hungry building: A review. *J. Build. Eng.* **2023**, *64*, 105604. [CrossRef]
13. Mesloub, A.; Ghosh, A.; Kolsi, L.; Alshenaifi, M. Polymer-dispersed liquid crystal (PDLC) smart switchable windows for less-energy hungry buildings and visual comfort in hot desert climate. *J. Build. Eng.* **2022**, *59*, 105101. [CrossRef]
14. Ghosh, A. Investigation of vacuum-integrated switchable polymer dispersed liquid crystal glazing for smart window application for less energy-hungry building. *Energy* **2023**, *265*, 126396. [CrossRef]
15. Ma, M.F.; Zhao, Y.Z.; Gao, H.; Wang, D.; Miao, Z.C.; Cao, H.; Yang, Z.; He, W.L. Role of semiconductor nanoparticles on electro-optical performance of PDLC devices. *Liq. Cryst.* **2022**, *49*, 2058–2068. [CrossRef]
16. Seo, J.; Song, M.; Jeong, J.; Nam, S.; Heo, I.; Park, S.Y.; Kang, I.K.; Lee, J.H.; Kim, H.; Kim, Y. Broadband pH-sensing organic transistors with polymeric sensing layers featuring liquid crystal microdomains encapsulated by di-block copolymer chains. *ACS Appl. Mater. Interfaces* **2016**, *8*, 23862–23867. [CrossRef] [PubMed]

17. Khlifi, S.; Bigeon, J.; Amela-Cortes, M.; Dumait, N.; Loas, G.; Cordier, S.; Molard, Y. Switchable two-dimensional waveguiding abilities of luminescent hybrid nanocomposites for active solar concentrators. *ACS Appl. Mater. Interfaces* **2020**, *12*, 14400–14407. [CrossRef]
18. Ailincai, D.; Marin, L. Eco-friendly PDLC composites based on chitosan and cholesteryl acetate. *J. Mol. Liq.* **2021**, *321*, 114466. [CrossRef]
19. Li, Y.; Gong, C.D.; Li, C.G.; Ruan, K.P.; Liu, C.; Liu, H.; Gu, J.W. Liquid crystalline texture and hydrogen bond on the thermal conductivities of intrinsic thermal conductive polymer films. *J. Mater. Sci. Technol.* **2021**, *82*, 250–256. [CrossRef]
20. Deng, Y.; Yang, Y.H.; Xiao, Y.H.; Xie, H.L.; Lan, R.C.; Zhang, L.Y.; Yang, H. Ultrafast switchable passive radiative cooling smart windows with synergistic optical modulation. *Adv. Funct. Mater.* **2023**, *33*, 2301319. [CrossRef]
21. Cheng, Z.X.; Wang, T.J.; Li, X.; Zhang, Y.H.; Yu, H.F. NIR-Vis-UV light-responsive actuator films of polymer-dispersed liquid crystal/graphene oxide nanocomposites. *ACS Appl. Mater. Interfaces* **2015**, *7*, 27494–27501. [CrossRef] [PubMed]
22. Zhang, P.; Tong, X.R.; Gao, Y.; Qian, Z.Y.; Ren, R.R.; Bian, C.C.; Wang, J.H.; Cai, G.F. A sensing and stretchable polymer-dispersed liquid crystal device based on spiderweb-inspired silver nanowires-micromesh transparent electrode. *Adv. Funct. Mater.* **2023**, *33*, 2303270. [CrossRef]
23. Shivaraja, S.J.; Gupta, R.K.; Manjuladevi, V. Faster switching polymer dispersed liquid crystal devices incorporated with functionalized SWCNTs. *J. Mol. Liq.* **2022**, *354*, 118905. [CrossRef]
24. Nasir, N.; Kumar, S.; Kim, M.; Nguyen, V.H.; Suleman, M.; Park, H.M.; Lee, S.; Kang, D.; Seo, Y. Effect of the photoinitiator concentration on the electro-optical properties of thiol-acrylate-based PDLC smart windows. *ACS Appl. Energy Mater.* **2022**, *5*, 6986–6995. [CrossRef]
25. Shi, Z.H.; He, Z.M.; Li, C.S.; Miao, Z.C.; Wang, D.; Luan, Y.; Li, Y.Z.; Zhao, Y.Z. The role of nanomesh fibres loaded with fluorescent materials on the electro-optical performance of PDLC devices. *Liq. Cryst.* **2022**, *49*, 2037–2050. [CrossRef]
26. Kim, E.; Liu, Y.; Hong, S.J.; Han, J.I. Effect of SiO_2 nanoparticle doping on electro-optical properties of polymer dispersed liquid crystal lens for smart electronic glasses. *Nano Converg.* **2015**, *2*, 5. [CrossRef]
27. Katariya-Jain, A.; Mhatre, M.M.; Dierking, I.; Deshmukh, R.R. Enhanced thermo-electro-optical and dielectric properties of carbon nanoparticle-doped polymer dispersed liquid crystal based switchable windows. *J. Mol. Liq.* **2024**, *393*, 123575. [CrossRef]
28. Zhang, Z.G.; Song, Y.Z.; Zhang, B.; Wang, L.; He, X.M. Metallized plastic foils: A Promising solution for high-energy lithium-ion battery current collectors. *Adv. Energy Mater.* **2023**, *13*, 2370146. [CrossRef]
29. Peng, S.J.; Du, X.F.; Liang, Z.S.; Ma, M.B.; Guo, Y.; Xiong, L.L. Multilayer polymer nanocomposites based on interface engineering for high-performance capacitors. *J. Energy Storage* **2023**, *60*, 106636. [CrossRef]
30. Khoi, T.M.; Tran, N.A.T.; Jung, H.B.; Huynh, V.P.; Kim, Y.; Hong, J.; Yoo, C.Y.; Kang, H.S.; Cho, Y. Selective and continuous ion recovery using flow electrode capacitive deionization with polymer multilayers functionalized ion exchange membrane. *Desalination* **2023**, *558*, 116626. [CrossRef]
31. Culebras, M.; Byun, Y.Y.; Jang, J.; Serafin, A.; Collins, M.N.; Park, Y.T.; Cho, C. Nanostructured PEDOT-based multilayer thin films with high thermoelectric performances. *Appl. Surf. Sci.* **2023**, *615*, 156432. [CrossRef]
32. Park, H.; Mong, A.L.; Kim, D. Single and multilayer composite electrolytes for enhanced Li-ion conductivity with restricted polysulfide diffusion for lithium-sulfur battery. *Mater. Today Energy* **2023**, *33*, 101274. [CrossRef]
33. Eslami, H.; Mekonnen, T.H. Flexible and green multilayer paper coating for barrier enhancement of paper packaging. *Sustain. Mater. Technol.* **2023**, *37*, e00694. [CrossRef]
34. Ogura, H.; Kawasaki, S.; Liu, Z.; Endo, T.; Maruyama, M.; Gao, Y.L.; Nakanishi, Y.; Lim, H.E.; Yanagi, K.; Irisawa, T.; et al. Multilayer in-plane heterostructures based on transition metal dichalcogenides for advanced electronics. *ACS Nano* **2023**, *17*, 6545–6554. [CrossRef]

molecules

MDPI

Article

Nanostructured Polymer-Dispersed Liquid Crystals Using a Ferroelectric Smectic A Liquid Crystal

Masaki Yamaguchi [1], Hiroyuki Matsukizono [2], Yasushi Okumura [2] and Hirotsugu Kikuchi [2,*]

[1] Interdisciplinary Graduate School of Engineering Sciences, Kyushu University, 6-1 Kasuga-Koen, Kasuga 816-8580, Fukuoka, Japan; yamaguchi.masaki.008@s.kyushu-u.ac.jp

[2] Institute for Materials Chemistry and Engineering, Kyushu University, 6-1 Kasuga-Koen, Kasuga 816-8580, Fukuoka, Japan; hmatsukizono@cm.kyushu-u.ac.jp (H.M.); okumura@cm.kyushu-u.ac.jp (Y.O.)

* Correspondence: kikuchi@cm.kyushu-u.ac.jp; Tel.: +81-92-583-7797

Abstract: Nanostructured polymer-dispersed liquid crystals (nano-PDLCs) are transparent and optically isotropic materials in which submicron-sized liquid crystal (LC) domains are dispersed within a polymer matrix. Nano-PDLCs can induce birefringence by applying an electric field (*E*-field) based on the reorientation of the LC molecules. If nano-PDLCs are utilized as light-scattering-less birefringence memory materials, it is necessary to suppress the relaxation of the LC molecule orientation after the removal of the *E*-field. We focused on the ferroelectric smectic A (SmA) phase to suppress the relaxation of LC molecules, owing to its layered structure and high viscosity. Although nano-PDLCs require a strong *E*-field to reorient their LC molecules because of the anchoring effect at the LC/polymer interface, the required field strength can be reduced using a ferroelectric smectic A (SmA_F) LC with a large dielectric constant. In this study, we fabricated a nano-PDLC by shining an ultraviolet light on a mixture comprised an SmA_F LC, photocurable monomers, and a photo-initiator. The electro-birefringence effect was evaluated using polarizing optical microscopy. After the removal of the *E*-field, an enhanced memory effect was observed in the sample using SmA_F LC compared with nematic LC-based nano-PDLCs.

Keywords: polymer-dispersed liquid crystals; memory effect; birefringence; molecular orientation

1. Introduction

Polymer-dispersed liquid crystals (PDLCs) are film-like solid composite materials comprising phase-separated liquid crystal (LC) domains and a polymer matrix. Polymerization-induced phase separation is the most common phase separation method. This method involves the irradiation of isotropic solutions of LCs, photocurable monomers, and a photo-initiator with ultraviolet (UV) light. The reorientation of LC molecules by applying an electric field (*E*-field) to PDLCs, which are electro-optical (EO)-responsive materials, has been used in practical applications [1–11]. PDLCs are generally opaque owing to their optical inhomogeneity, which causes transmitted light to scatter. Applying an *E*-field to a PDLC renders it optically uniform and transparent owing to its LC reorientation. Additionally, PDLCs exhibit light scattering–light transmission switching upon turning the *E*-field off and on, respectively. This characteristic facilitates their application as light-regulating materials. Transparent PDLCs (nano-PDLCs), even in the absence of an applied *E*-field, can be obtained by reducing the LC weight fraction in the raw material while increasing the UV irradiation intensity to form a phase-separated structure that is smaller than a visible wavelength. Nano-PDLCs, such as those with suppressed random light scattering, have been proposed as attractive EO-responsive materials for holographic films [12], microlenses [13], phase modulators [14,15], and display elements without view–angle dependence based on the electro-optical Kerr effect [16–23]. While general PDLCs switch between light scattering and transmission with the *E*-field off and on, respectively, nano-PDLCs are transparent materials regardless of whether an electric field is applied and are initially optically isotropic because of the random orientation of the LC molecules. However, when an *E*-field is applied, LCs' reorientation and macroscopic birefringence are induced, resulting in an electro-birefringence effect. The EO responsivity of nano-PDLCs is also based on the reorientation of LC molecules inside the LC domain by an applied *E*-field. Because of the smaller size of the LC domains, nano-PDLCs require a stronger *E*-field than general PDLCs for their driving because of the large specific surface area of the LC/polymer interface and the strong influence of interface anchoring. High voltages are required to drive nano-PDLCs. Therefore, attempts have been made to lower the driving voltage by doping nano-PDLCs with conductive materials [20–22], such as low-T_g polymers [23]. Recently, we developed a transparent PDLC using a highly polar nematic (N) LC mixture, an LC molecule bearing a 1,3-dioxane skeleton (DIO) [24], and its analogs. These compounds exhibit large dielectric constant (ε') values and induce birefringence at low driving voltages. The induced birefringence was partially maintained even after the removal of the *E*-field (memory effect) [25]. In a previous study, the birefringence viewed from the normal direction of the substrate was reversibly erased by switching the electrical circuit with a relay switch and applying an *E*-field in the out-of-plane direction of the substrate. Nano-PDLCs exhibit an electro-birefringence effect based on the reorientation of LC molecules; memorized birefringence (up to 50% of the induced birefringence) is also derived from the retention of their molecular orientation. The orientation memory effect of LC molecules after the removal of the *E*-field is expected to be more stable in layered structures with mechanical stability and in highly viscous smectic (Sm) LC phases; consequently, Sm LC-based memory-type PDLCs have also been reported [26,27]. In general, LC molecules in the smectic A (SmA) phase form a layered structure and are more viscous than those in the N phase. Therefore, the threshold voltage required for reorientation by the *E*-field of the SmA phase is higher than that of the N phase. However, the high elastic modulus also suppressed the relaxation of the molecular orientation after removal of the *E*-field. Therefore, we focus on SmA LCs with high dielectric constants parallel to the LC director. It has been demonstrated that specific DIO analog molecules exhibit a ferroelectric SmA (SmA_F) phase with spontaneous polarization parallel to the LC director [28–32]. The relative dielectric constant of SmA_F LCs can reach several hundred [28,29,32]. Materials that can be driven at voltages lower than the driving voltages of conventional SmA LC-based materials while exhibiting a better retention of their molecular orientation than N LCs can be developed using SmA LC materials with greater dielectric anisotropy than that of conventional materials. Our earlier study showed that, in comparison to conventional SmA LCs, LC molecules with an ester skeleton had higher dielectric constant values (EST-4) [32]. In this study, we fabricated a transparent PDLC using an SmA_F LC (hereafter referred to as an EST). The structural and physical properties of the fabricated PDLC was assessed, and the impact of memory on the electro-birefringence effect was investigated.

2. Results and Discussion

At temperatures over 80 °C, where EST exhibited LC phases, the PDLC exhibited a higher normalized transmittance ($T > 0.7$), suggesting the formation of phase-separated structures smaller than the visible light wavelength (Figure 1). Furthermore, T decreases with decreasing temperature. This decrease in T can be attributed to the slight coarsening of the phase-separated LC domains due to the lowering of the compatibility between the EST molecules and the polymer matrix with decreasing temperature. Scanning electron microscopy (SEM) images show a phase separation of approximately 60–120 nm with a polymer ball-type morphology (Figure 1 inset and Figure S1 in the Supplementary Materials). LC droplet structures have not been observed in memory PDLCs with micrometer-sized phase-separated structures [33]. However, a similar polymer morphology was observed in the present study.

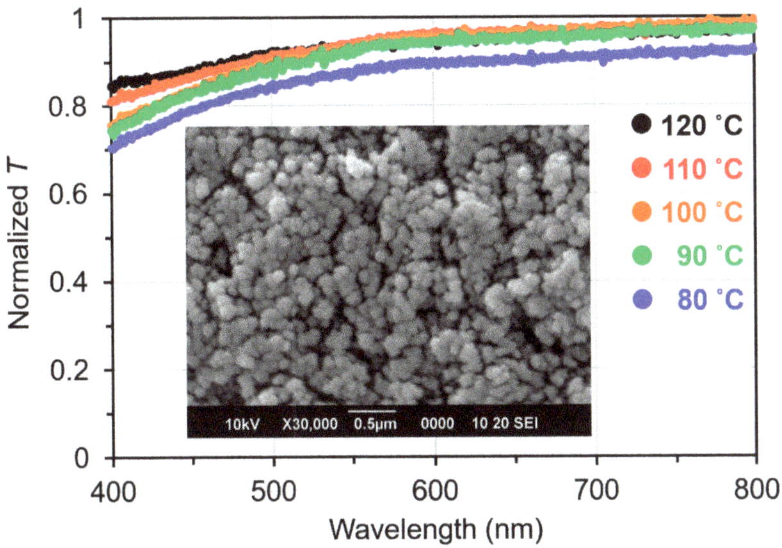

Figure 1. Transmittance of a PDLC normalized with a toluene-filled cell. Inset: SEM image of the PDLC of the polymer matrix.

All the observed phase transition points during heating and cooling shifted toward lower temperatures than those of the bulk EST [32]. During the cooling step, exothermic peaks were observed at 130 °C (Iso. to N), 111 °C (N to SmA_F), and 74 °C (SmA_F to Cr), depending on the phase transition (Figure 2a). Furthermore, the phase transition enthalpy of EST in the PDLC was lower than that of the bulk—0.328, 1.18, and 15.1 kJ/mol, which were 78, 45, and 56% of that of the bulk, respectively. The small apparent enthalpies of the phase transitions observed in the differential scanning calorimetry (DSC) measurements can be attributed to the fact that many of the ESTs dissolved in the polymer matrix or were dispersed to such a small size that they did not exhibit a phase transition. The lower transition point of Iso. to N in the PDLC compared to the bulk LC suggests a higher solubility or affinity between the LC and the polymer [34]. Furthermore, the broadening of each phase transition peak in the DSC curves suggests that the materials undergoing phase transitions are thermally destabilized and that their degree of destabilization varies widely. The phase transitions of EST occur over a broad temperature range, indicating that EST molecules adopt various states in the polymer matrix.

Next, dielectric measurements were performed to investigate the relationship between the phase transitions and the electrical properties of the PDLC. The dielectric constant (ε') of the PDLC was measured within the range of 130–70 °C during the cooling process, as shown in Figure 2b. A comparison of the temperature dependence of ε' at different frequencies indicates that the magnitude of ε' tends to be smaller at higher frequencies, with a large change in ε' between 115 and 110 °C at all frequencies. This marked temperature-dependent change in ε' may be associated with the change in *E*-field responsivity due to the SmA_F-to-N phase transition of the dispersed EST molecules in the polymer matrix. The ε' value is large in the low-frequency range; however, the large apparent ε' may include the conductive components of small amounts of adsorbed water in the material and small amounts of ionic conductive impurities. In addition, LC domains in an insulating polymer matrix are less likely to be subjected to an E-field, which should be considered in the low-frequency range [35]. As evident from the DSC curves in Figure 2a, the change in electrical properties at approximately 115–110 °C can be attributed to the phase transition of EST molecules.

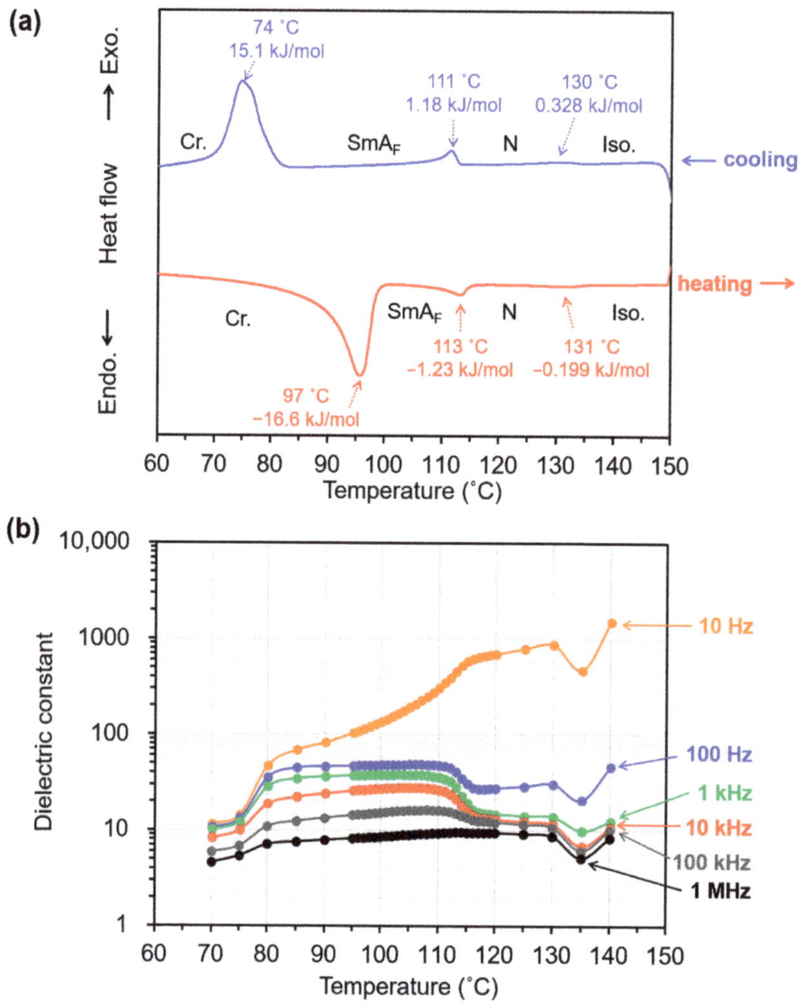

Figure 2. (**a**) DSC curves of the PDLC; scanning rate: 5 °C/min. (**b**) Temperature dependence of the dielectric constant (ε') of the PDLC at various frequencies.

As regards Δn values, Δn_{app}. and Δn_{rem}. denote the Δn values during the application of an *E*-field and after the removal of an *E*-field, respectively. As shown in Figure 3a,d,g, before the application of the *E*-field, the PDLC sample does not show a Δn and does not transmit light in the polarizing optical microscopy (POM) observation under crossed nicols. When a voltage of 100 V was applied at a frequency of 10 kHz, Δn_{app}. was induced by the *E*-field because light was transmitted between the electrodes (Figure 3b,e,h). The same procedure was also used to measure Δn_{app}. at frequencies of 1 kHz, 100 Hz, and 10 Hz and at various temperatures when 100 V was applied. A comparison of Δn_{app}. during the application of 100 V at various frequencies indicates that a larger Δn_{app}. was observed at higher frequencies (Figure 3j). This is contrary to the trend in ε', wherein larger ε' values were observed at lower frequencies. This result suggests that when 100 V is applied to the PDLC, a sufficient voltage is not applied to the EST-rich domain to reorient the EST molecules, unless it is in the high-frequency range of approximately 1 kHz or higher. If larger phase-separated structures were obtained, the PDLC would exhibit higher light scattering; however, a larger $\Delta n_{app.}$ would be induced. The $\Delta n_{app.}$ was almost zero during the application of 100 V in the case of a transparent PDLC with an SmA LC, 4-cyano-4′-*n*-octyloxybiphenyl (8OCB). This also confirms that PDLCs using EST are more EO-responsive than other SmA LC-based materials. After the removal of the *E*-field, $\Delta n_{rem.}$ decreased significantly, as only a small amount of light was transmitted between the electrodes at 115 °C (Figure 3c). Meanwhile, at 110 and 100 °C, the light intensity transmitted between the electrodes was higher (brighter), indicating the retention of larger Δn_{rem} values (Figure 3f,i). The retention of $\Delta n_{rem.}$ after the removal of the *E*-field indicates that the EST molecules did not completely relax to their initial (random) molecular orientation after the removal of the *E*-field and that the molecular orientation induced by the *E*-field was partially maintained. $\Delta n_{rem.}$ was completely erased upon heating the sample above 130 °C.

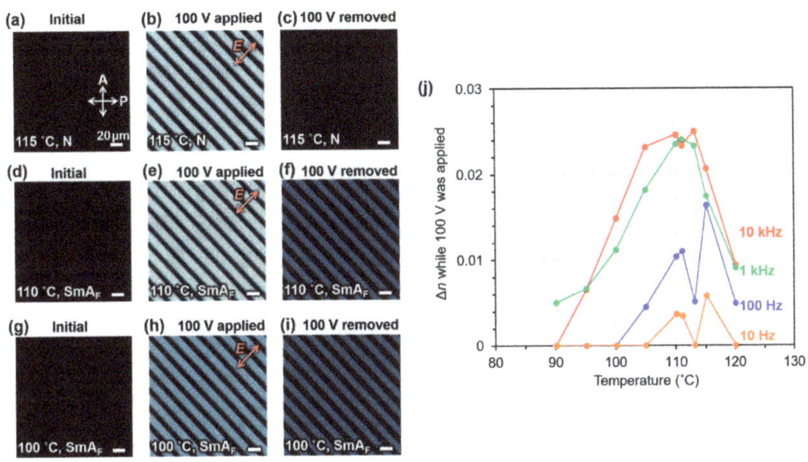

Figure 3. (**a**,**d**,**g**) POM images of the PDLC before applying an *E*-field at 115, 110, and 100 °C. (**b**,**e**,**h**) POM images of the PDLC during application of 100 V with a frequency of 10 kHz at 115, 110, and 100 °C. (**c**,**f**,**i**) POM images of the PDLC after removing the *E*-field at 115, 110, and 100 °C. (**j**) Temperature dependence of *E*-field-induced Δn during the application of 100 V at various frequencies.

The relationship between the Δn memory properties and temperatures after the removal of the *E*-field was meticulously investigated for the frequencies of 1 and 10 kHz, where better electro-birefringence effects were observed. The proportion of Δn memorized after the removal of the *E*-field was defined as the memory retention rate, MRR, which is calculated as follows:

$$\mathrm{MRR}\ (\%) \equiv \frac{\Delta n_{rem.}}{\Delta n_{app.}} \times 100. \tag{1}$$

The temperature dependence of the MRR was less than 40% above 111 °C, indicating that EST was the N phase in the PDLC sample. However, below 100 °C, where EST is considered to have completely transitioned to SmA$_F$, the MRR was more than 60%. These features were similar at the frequencies of 1 and 10 kHz (Figure 4). As evident from the DSC curve in Figure 2a, the phase transition from N to SmA$_F$ proceeds within broad temperature ranges; the MRR was not likely to be high given the presence of both the SmA$_F$ phase and N phase at 110 °C. At higher temperatures, the MRR was less than 40% because of the relaxation of the molecular orientation of EST to its initial orientation upon the removal of the *E*-field. In contrast, in the temperature range of the SmA$_F$ phase, the elastic modulus of EST was higher than that of the N phase, and the relaxation of its molecular orientation after the removal of the *E*-field was suppressed. In addition, by using the positive-up-negative-down (PUND) method [36], currents due to polarization reversal were observed, confirming its ferroelectricity (Figure S2 in the Supplementary Materials).

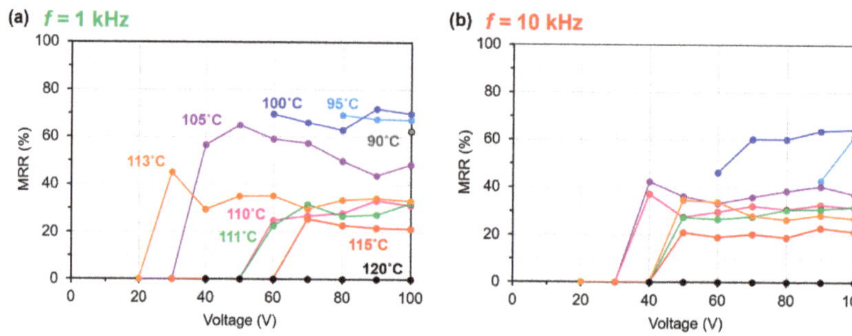

Figure 4. MRR in the temperature range 120–90 °C at the frequency of (**a**)1 kHz and (**b**) 10 kHz.

Previously, we fabricated N LC-based nano-PDLCs with phase-separated structure sizes of 100–200 nm and a maximum MRR of 50% [25]. The polar monofunctional vinyl monomer NVP, which is thought to provide a polar anchoring force between the polymer interface and EST molecules, was used in this study. The anchoring effect was expected to be stronger in the nano-PDLC in this study than in the previous study because of the lower LC-constitutive fraction and smaller phase-separated structure. Furthermore, the previous work did not use NVP as the monomer. Because anchoring at the interface with the polymer is the driving source of LC reorientation after the removal of the *E*-field, a strong anchoring effect is disadvantageous for the memory effect. The high MRR (70%) in this study, despite a stronger anchoring than that in previous studies, indicates that the ability of SmA_F LCs to retain their molecular orientation after *E*-field removal is beneficial. Because the LC domains within the nano-PDLCs were remarkably fine, the LCs confined within them were subjected to large deformations. In the memory state of nano-PDLCs, slight deformations of the Sm layers and the presence of minor defects may be acceptable. Conversely, large layer deformations and numerous defects can lead to memory degradation [27]. The loss of approximately 30% of the Δn memory is apparently due to the complex structure of the polymer and anchoring forces that distort the molecular orientation of EST near the polymer interface. This results in the inability to maintain an Sm-layered structure near the polymer, as well as partial N-like orientation ordering and enhanced relaxation. The anchoring of polymers with large polarities may be effective in maintaining their molecular orientation by retaining the ferroelectric polarization of their EST molecules. The PDLC fabricated in this study can be driven at voltages lower than those of conventional SmA LC-based materials. It can also be pinned at any Δn value more efficiently than N LC-based materials, making it suitable for applications such as display elements without viewing angle dependence and electrically tunable microlenses.

3. Materials and Methods

3.1. PDLC Sample Preparation

The PDLC precursor was prepared by mixing a 50 wt% (2.73 equiv.) EST as the SmA_F LC material (synthesized in a previous study [32]); 24.5 wt% (1.0 equiv.) Dipentaerythritol hexaacrylate (DPEHA; Tokyo Chemical Industry Co., Ltd., Tokyo, Japan) and 24.5 wt% (5.21 equiv.) of *N*-vinyl-2-pyrrolidone (NVP, Tokyo Chemical Industry Co., Ltd.) as photo-polymerizable monomers; and 1 wt% (9.21×10^{-2} equiv.) 2,2-dimethoxy-2-phenylacetophenone (DMPAP, Tokyo Chemical Industry Co., Ltd.) as a photo-initiator. The chemical structures of the PDLC precursors are shown in Figure 5. The hexa-functional acrylic monomer, DPEHA, was employed for curing during the initial stages of polymerization-induced phase separation. NVP, a polar vinyl monomer, was used as a solubilizer for EST with a high melting point (100 °C in bulk). The PDLC precursor was injected by capillary action into an indium tin oxide (ITO)-patterned glass cell (ISSZ-10/B707M7NSS, E.H.C. Co., Ltd., Tokyo, Japan) at 100 °C in an isotropic solution.

UV light (λ = 365 nm peak, 50 mW/cm^2) was then applied for 5 min at 110 °C for the photo-polymerization and phase separation to progress, yielding a transparent PDLC.

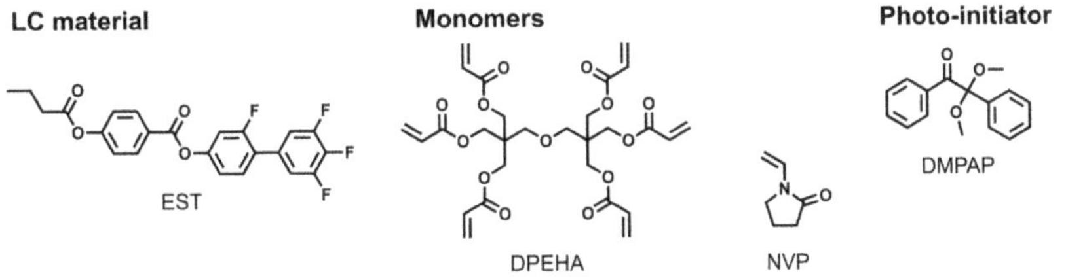

Figure 5. Chemical structure of substances in the PDLC precursor.

3.2. Light Scattering Characterization of PDLC

To investigate the structure of the resulting PDLC, a light scattering characterization of the material and the transmitted light intensity was performed using a microscope (Axio Imager.A2M, ZEISS, Oberkochen, Germany) and a compact instantaneous spectroscopic measurement unit (SA-100S-CK1, LAMBDA VISION Inc., Yokohama, Japan). The normalized transmittance (T) was calculated by measuring the transmitted light intensity in the wavelength range of 400–800 nm and normalizing it to the transmitted light intensity of a toluene-filled cell.

3.3. Scanning Electron Microscopy (SEM) Observation of Polymer Matrix

The morphology of the polymer matrix was observed using SEM. For SEM observations, the PDLC sample was soaked in dichloromethane to extract the EST molecules and dried thoroughly under vacuum. Finally, the polymer matrix was sputtered with Pt, and its morphology was characterized using SEM (Carry Scope JCM5700, JEOL Co., Ltd., Tokyo, Japan) at an accelerating voltage of 10 kV.

3.4. Differential Scanning Calorimetry (DSC) Measurements

Differential scanning calorimetry (DSC) of the PDLC sample was performed to investigate the phase transition behavior of EST in a fine polymer matrix. DSC curves were recorded using a differential scanning calorimeter (DSC 1 STAR^e System, Mettler Toledo, Greifensee, Switzerland) with a dedicated Al pan at a scanning rate of 5 °C/min.

3.5. Dielectric Spectroscopy of PDLC

The dielectric relaxation spectra of the PDLC were recorded in the range of 1 Hz to 11 MHz using an impedance/gain phase analyzer (SI 1260, Solartron Metrology, Bognor Regis, UK) at an applied voltage of 0.1 V. Cells were used with no surface orientation treatment, an ITO electrode area of 1 cm^2, and a cell thickness of 10 µm (KSSZ-10/B107M6NSS05, E.H.C. Co., Ltd.). First, the ITO electrode's resistance and capacitance were measured using an empty cell to correct the PDLC impedance and obtain the PDLC dielectric constants. After measuring the empty cell's resistance and capacitance, the precursor was injected into the cell and photo-polymerized under UV irradiation.

3.6. Electro-Birefringence Effect

The electro-birefringence effect of the PDLC sample was measured using a polarizing optical microscope (ECLIPSE LV100 POL, Nikon, Tokyo, Japan) with a DS-Ri1 camera under crossed polarizers and by applying a sine-wave E-field parallel to the substrate plane with frequencies ranging from 10 Hz to 10 kHz. To determine the frequency characteristics of the electro-birefringence effect of the PDLC sample, the optical retardation (at 536 nm) of each sample was measured using a Berek compensator (Nichika Co., Ltd., Kobe, Japan). The birefringence (Δn) was calculated by dividing the optical retardation by the cell thickness (10 µm).

4. Conclusions

A PDLC exhibiting high transparency in the visible wavelength range was fabricated using SmA_F LC, photopolymerizable monomers, and a photo-initiator. SEM observations revealed the formation of a fine polymer ball morphology with a size of 60–120 nm; no LC droplet morphology was observed. The phase transition behavior of the LC molecules in the PDLC was investigated using DSC, and the results suggested strong interactions between the polymer matrix and the LC molecules. Furthermore, with regard to the electro-birefringence effect, different electric birefringence responses were observed depending on the phase transition of the LCs in the PDLC. After the LCs' transition from the N phase to the SmA_F phase, the proportion of memorized Δn retained after the removal of the E-field increased, suggesting enhanced molecular orientation memory based on the mechanical stability in the SmA_F phase. These can be driven at voltages lower than those of conventional materials using SmA LCs. In addition, the molecular orientation memory, which was enhanced in the SmA_F phase, was stable after the removal of the E-field. These PDLCs can potentially aid the development of applications such as display elements without view–angle dependence or electrically tunable microlenses.

Supplementary Materials: The following supporting information can be downloaded at https://www.mdpi.com/article/10.3390/molecules29204837/s1, Figure S1: SEM images of polymer matrix at different magnifications.; Figure S2: (a) Switching current response of the PDLC while applying a triangular-wave E-field. (b) Hysteresis between the electric flux density and voltage measured in the SmA_F phase, see ref. [36].

Author Contributions: Conceptualization, H.K.; methodology, H.K.; validation, M.Y.; formal analysis, M.Y.; investigation, M.Y., H.M. and Y.O.; resources, H.M., Y.O. and H.K.; data curation, M.Y.; writing—original draft preparation, M.Y.; writing—review and editing, H.M., Y.O. and H.K.; visualization, M.Y.; supervision, H.K.; project administration, H.K.; funding acquisition, H.K. All authors have read and agreed to the published version of the manuscript.

Funding: This research was funded by JSPS KAKENHI, Grant Numbers JP23H00303 and JP23K17366; MEXT Project "Integrated Research Consortium on Chemical Sciences (IRCCS)", Dynamic Alliance for Open Innovation Bridging Human, Environment and Materials from the Ministry of Education, Culture, Sports, Science and Technology, Japan (MEXT); and the Cooperative Research Program of the "Network Joint Research Center for Materials and Device".

Institutional Review Board Statement: Not applicable.

Informed Consent Statement: Not applicable.

Data Availability Statement: Data are contained within the article.

Conflicts of Interest: The authors declare no conflicts of interest.

References

1. Kajiyama, T.; Miyamoto, A.; Kikuchi, H.; Morimura, Y. Aggregation States and Electro-optical Properties Based on Light Scattering of Polymer/(Liquid Crystal) Composite Films. *Chem. Lett.* **1989**, *18*, 813–816. [CrossRef]
2. Kajiyama, T.; Kikuchi, H.; Miyamoto, A.; Moritomi, S.; Hwang, J. Aggregation States and Bistable Light Switching of (Liquid Crystalline Polymer)/(Low Molecular Weight Liquid Crystal) Mixture Systems. *Chem. Lett.* **1989**, *18*, 817–820. [CrossRef]
3. Doane, J.W.; Golemme, A.; West, J.L.; Whitehead, J.B.; Wu, B.G. Polymer Dispersed Liquid Crystals for Display Application. *Mol. Cryst. Liq. Cryst.* **1988**, *165*, 511–532. [CrossRef]
4. Hikmet, R.A.M.; Kemperman, H. Electrically switchable mirrors and optical components made from liquid-crystal gels. *Nature* **1998**, *392*, 476–479. [CrossRef]
5. Wu, B.G.; Erdmann, J.H.; Doane, J.W. Response times and voltages for PDLC light shutters. *Liq. Cryst.* **1989**, *5*, 1453–1465. [CrossRef]
6. Sheraw, C.D.; Zhou, L.; Huang, J.R.; Gundlach, D.J.; Jackson, T.N.; Kane, M.G.; Hill, I.G.; Hammond, M.S.; Campi, J.; Greening, B.K.; et al. Organic thin-film transistor-driven polymer-dispersed liquid crystal displays on flexible polymeric substrates. *Appl. Phys. Lett.* **2002**, *80*, 1088–1090. [CrossRef]
7. Geis, M.W.; Lyszczarz, T.M.; Osgood, R.M.; Kimball, B.R. 30 to 50 ns liquid-crystal optical switches. *Opt. Express* **2010**, *18*, 18886–18893. [CrossRef]

8. Gong, S.; Cao, Y.; Fang, X.; Wang, Y.F.; Liu, Q.Z.; Gui, H.; Shen, C.F.; Cao, X.; Kim, E.S.; Zhou, C.W. Carbon Nanotube Macroelectronics for Active Matrix Polymer-Dispersed Liquid Crystal Displays. *ACS Nano* **2016**, *10*, 10068–10074. [CrossRef]
9. Oh, S.W.; Baek, J.M.; Heo, J.; Yoon, T.H. Dye-doped cholesteric liquid crystal light shutter with a polymer-dispersed liquid crystal film. *Dye. Pigment.* **2016**, *134*, 36–40. [CrossRef]
10. Chen, G.N.; Ni, M.L.; Peng, H.Y.; Huang, F.H.; Liao, Y.G.; Wane, M.K.; Zhu, J.T.; Roy, V.A.L.; Xie, X.L. Photoinitiation and Inhibition under Monochromatic Green Light for Storage of Colored 3D Images in Holographic Polymer-Dispersed Liquid Crystals. *ACS Appl. Mater. Interfaces* **2017**, *9*, 1810–1819. [CrossRef]
11. Wang, K.N.; Zheng, J.H.; Liu, Y.R.; Gao, H.; Zhuang, S.L. Electrically tunable two-dimensional holographic polymer-dispersed liquid crystal grating with variable period. *Opt. Commun.* **2017**, *392*, 128–134. [CrossRef]
12. Bunning, T.J.; Natarajan, L.V.; Tondiglia, V.P.; Sutherland, R.L. Holographic polymer-dispersed liquid crystals (H-PDLCs). *Annu. Rev. Mater. Sci.* **2000**, *30*, 83–115. [CrossRef]
13. Pagidi, S.; Manda, R.; Bhattacharyya, S.S.; Lee, S.G.; Song, S.M.; Lim, Y.J.; Lee, J.H.; Lee, S.H. Fast Switchable Micro-Lenticular Lens Arrays Using Highly Transparent Nano-Polymer Dispersed Liquid Crystals. *Adv. Mater. Interfaces* **2019**, *6*, 1900841. [CrossRef]
14. Verbrugge, V.; de la Tocnaye, J.; Dupont, L. C-band wavelength-tunable vertical-cavity laser using a nano polymer dispersed liquid crystal material. *Opt. Commun.* **2003**, *215*, 353–359. [CrossRef]
15. Levallois, C.; Verbrugge, V.; Dupont, L.; de la Tocnaye, J.L.B.; Caillaud, B.; Le Correa, A.; Dehaese, O.; Folliot, H.; Loualiche, S. 1.55-μm optically pumped tunable VCSEL based on a nano-polymer dispersive liquid crystal phase modulator. In Proceedings of the Conference on Micro-Optics, VCSELs, and Photonic Interconnects II, Strasbourg, France, 3–5 April 2006. [CrossRef]
16. Aya, S.; Le, K.V.; Araoka, F.; Ishikawa, K.; Takezoe, H. Nanosize-Induced Optically Isotropic Nematic Phase. *Jpn. J. Appl. Phys.* **2011**, *50*, 051703. [CrossRef]
17. Matsumoto, S.; Houlbert, M.; Hayashi, T.; Kubodera, K. Fine droplets of liquid crystals in a transparent polymer and their response to an electric field. *Appl. Phys. Lett.* **1996**, *69*, 1044–1046. [CrossRef]
18. Cho, N.H.; Nayek, P.; Lee, J.J.; Lim, Y.J.; Lee, J.H.; Lee, S.H.; Park, H.S.; Lee, H.J.; Kim, H.S. High-performance, in-plane switching liquid crystal device utilizing an optically isotropic liquid crystal blend of nanostructured liquid crystal droplets in a polymer matrix. *Mater. Lett.* **2015**, *153*, 136–139. [CrossRef]
19. Lim, Y.J.; Yoon, J.H.; Yoo, H.; Song, S.M.; Manda, R.; Pagidi, S.; Lee, M.H.; Myoung, J.M.; Lee, S.H. Fast switchable field-induced optical birefringence in highly transparent polymer-liquid crystal composite. *Opt. Mater. Express* **2018**, *8*, 3698–3707. [CrossRef]
20. Pagidi, S.; Manda, R.; Bhattacharyya, S.S.; Cho, K.J.; Kim, T.H.; Lim, Y.J.; Lee, S.H. Superior electro-optics of nano-phase encapsulated liquid crystals utilizing functionalized carbon nanotubes. *Compos. Part B-Eng.* **2019**, *164*, 675–682. [CrossRef]
21. Pagidi, S.; Manda, R.; Shin, H.S.; Lee, J.; Lim, Y.J.; Kim, M.; Lee, S.H. Enhanced electro-optic characteristics of polymer-dispersed nano-sized liquid crystal droplets utilizing PEDOT:PSS polymer composite. *J. Mol. Liq.* **2021**, *322*, 114959. [CrossRef]
22. Pagidi, S.; Srivastava, A.; Pandey, N.; Manda, R. Reduced driving electric field of ionic salt doped nano-structured polymer dispersed liquid crystals device. *J. Mol. Liq.* **2024**, *399*, 124444. [CrossRef]
23. Pagidi, S.; Park, H.; Lee, D.; Kim, M.; Lee, S.H. Nanosize-confined nematic liquid crystals at slippery interfaces of polymer composites consisting of poly (hexyl methacrylate). *J. Mol. Liq.* **2022**, *350*, 118540. [CrossRef]
24. Nishikawa, H.; Shiroshita, K.; Higuchi, H.; Okumura, Y.; Haseba, Y.; Yamamoto, S.I.; Sago, K.; Kikuchi, H. A Fluid Liquid-Crystal Material with Highly Polar Order. *Adv. Mater.* **2017**, *29*, 1702354. [CrossRef]
25. Yamaguchi, M.; Okumura, Y.; Nishikawa, H.; Matsukizono, H.; Kikuchi, H. Memorizable Electro-Birefringence Effect Exhibited by Transparent Liquid Crystal/Polymer Composite Materials. *Adv. Electron. Mater.* **2024**, *10*, 2400055. [CrossRef]
26. Fuh, A.; Ko, T.; Li, M. Polymer Dispersed Liquid Crystal Films with Memory Characteristics. *Jpn. J. Appl. Phys.* **1992**, *31*, 3366–3369. [CrossRef]
27. Date, M.; Takeuchi, Y.; Kato, K. Droplet size effect on the memory-mode operating temperature of smectic-A holographic polymer dispersed liquid crystal. *J. Phys. D Appl. Phys.* **1999**, *32*, 3164–3168. [CrossRef]
28. Kikuchi, H.; Matsukizono, H.; Iwamatsu, K.; Endo, S.; Anan, S.; Okumura, Y. Fluid Layered Ferroelectrics with Global $C_{\infty v}$ Symmetry. *Adv. Sci.* **2022**, *9*, 2202048. [CrossRef]
29. Song, Y.; Deng, M.; Wang, Z.; Li, J.; Lei, H.; Wan, Z.; Xia, R.; Aya, S.; Huang, M. Emerging Ferroelectric Uniaxial Lamellar (Smectic A_F) Fluids for Bistable In-Plane Polarization Memory. *J. Phys. Chem. Lett.* **2022**, *13*, 9983–9990. [CrossRef] [PubMed]
30. Chen, X.; Martinez, V.; Nacke, P.; Korblova, E.; Manabe, A.; Klasen-Memmer, M.; Freychet, G.; Zhernenkov, M.; Glaser, M.; Radzihovsky, L.; et al. Observation of a uniaxial ferroelectric smectic A phase. *Proc. Natl. Acad. Sci. USA* **2022**, *119*, e2210062119. [CrossRef]
31. Gibb, C.; Hobbs, J.; Nikolova, D.; Raistrick, T.; Berrow, S.; Mertelj, A.; Osterman, N.; Sebastian, N.; Gleeson, H.; Mandle, R. Spontaneous symmetry breaking in polar fluids. *Nat. Commun.* **2024**, *15*, 5845. [CrossRef]
32. Matsukizono, H.; Sakamoto, Y.; Okumura, Y.; Kikuchi, H. Exploring the Impact of Linkage Structure in Ferroelectric Nematic and Smectic Liquid Crystals. *J. Phys. Chem. Lett.* **2024**, *15*, 4212–4217. [CrossRef] [PubMed]
33. Yamaguchi, R.; Sato, S. Memory Effects of Light Transmission Properties in Polymer-Dispersed-Liquid-Crystal (PDLC) Films. *Jpn. J. Appl. Phys.* **1991**, *30*, L616–L618. [CrossRef]
34. Mucha, M. Polymer as an important component of blends and composites with liquid crystals. *Prog. Polym. Sci.* **2003**, *28*, 837–873. [CrossRef]

35. Miyamoto, A.; Kikuchi, H.; Kobayashi, S.; Morimura, Y.; Kajiyama, T. Dielectric property-electrooptical effect relationships of polymer/liquid-crystal composite films. *Macromolecules* **1991**, *24*, 3915–3920. [CrossRef]
36. Fukunaga, M.; Noda, Y. New technique for measuring ferroelectric and antiferroelectric hysteresis loops. *J. Phys. Soc. Jpn.* **2008**, *77*, 064706. [CrossRef]

molecules

MDPI

Article

Synthesis of Side-Chain Liquid Crystalline Polyacrylates with Bridged Stilbene Mesogens

Gen-ichi Konishi [1,2,*], Yuki Sawatari [1], Riki Iwai [1], Takuya Tanaka [1], Yoshimichi Shimomura [1] and Masatoshi Tokita [1,2]

[1] Department of Chemical Science and Engineering, Institute of Science Tokyo, Tokyo 152-8552, Japan
[2] Department of Polymer Chemistry, Tokyo Institute of Technology, Tokyo 152-8552, Japan
* Correspondence: konishi.g.579e@m.isct.ac.jp or konishi.g.aa@m.titech.ac.jp

Abstract: In recent years, π-conjugated liquid crystalline molecules with optoelectronic functionalities have garnered considerable attention, and integrating these molecules into side-chain liquid crystalline polymers (SCLCPs) holds potential for developing devices that are operational near room temperature. However, it is difficult to design SCLCPs with excellent processability because liquid crystalline mesogens are rigid rods, have low solubility in organic solvents, and have a high isotropization temperature. Recently, we developed near-room-temperature π-conjugated nematic liquid crystals based on "bridged stilbene". In this work, we synthesized a polyacrylate SCLCP incorporating a bridged stilbene that exhibited a nematic phase near room temperature and could maintain liquid crystallinity for more than three months. We conducted a thorough phase structure analysis and evaluated the optical properties. The birefringence values of the resulting polymers were higher than those of the corresponding monomers because of the enhanced order parameters due to the polymer effect. In addition, the synthesized polymers inherited mesogen-derived AIE properties, with high quantum yields (Φ_{fl} = 0.14–0.35) in the solid state. It is noteworthy that the maximum fluorescence wavelength exhibited a redshift of greater than 27 nm as a consequence of film formation. Thus, several unique characteristics of the SCLCPs are unattainable with small molecular systems.

Keywords: side-chain liquid crystalline polymer; nematic liquid crystal; polyacrylate; π-conjugated mesogen; birefringence; aggregation-induced emission

1. Introduction

Liquid crystal polymers are widely used in industrial applications due to their excellent processability and favorable material properties [1–14]. Of particular interest are side-chain liquid crystal polymers (SCLCPs) [15–25], in which mesogenic units are attached to the side chains of linear polymers such as polyacrylate and polymethacrylate. SCLCPs exhibit liquid crystalline phases over a broader temperature range than small molecule liquid crystals and mimic the behavior of small molecule liquid crystals above the glass transition temperature (T_g). This liquid crystalline nature allows for the molecular orientation to be controlled by external fields such as mechanical stress, electric fields, and magnetic fields [1–14]. In addition, the molecular orientation of LC polymers can be controlled using anisotropic surfaces (as in this study) and photoalignment (e.g., [16]). Additionally, SCLCPs can form nanostructures through higher order smectic and columnar phases [26–29]. Below T_g, the liquid crystal phase that exists above T_g can be vitrified, retaining its anisotropic properties. Among liquid crystal phases, the nematic (N) phase is the most fluid, is highly responsive to external fields, and is suitable for large-area applications [30]. These properties, unique to the N phase, distinguish it from other liquid crystal phases and extend its range of applications beyond traditional uses, including displays [31,32] and polarizing films, into advanced optical materials. The properties exhibited by SCLCPs are highly attractive for the development of materials that leverage the unique optical, luminescent, and electronic properties of π-conjugated systems [33–46]. In particular, SCLCPs with liquid crystalline organic semiconductors in the side chains can be one of the most powerful tools for fabricating film-like devices with bulk arrays of organic semiconductors. In other words, it is effective for the realization of advanced materials based on functional liquid crystalline π-conjugated molecules, such as displays and molecular electronics. However, π-conjugated molecules with functional groups typically have high melting points, and SCLCPs incorporating these as mesogens often struggle to exhibit the nematic (N) phase or only exhibit it at elevated temperatures above 100 °C [45]. In addition, the introduction of a long-chain alkyl group into the mesogen lowers the isotropization temperature but tends to produce a smectic phase with high crystallinity and low operability [47]. As a result, it is challenging to achieve room temperature N-phase behavior in SCLCPs that feature π-conjugated mesogens with photo/electronic functionality.

Recently, we developed a novel aggregation-induced emission luminogen (AIEgen) [48–54], "seven-membered bridged stilbene", based on 4-phenyl stilbene (**PST**) with a seven-membered ring structure, specifically 3,8-diphenyl-6,7-dihydro-5*H*-benzo[7]annulene (**DPB[7]**) [55–57]. We reported that **DPB[7]**, when modified with appropriate alkyl tails at both ends, exhibits the N phase around room temperature [58]. This bridged stilbene liquid crystal (LC) possesses the unique ability to maintain the N phase over a wide temperature range while exhibiting luminescent properties. Such π-conjugated liquid crystals with low isotropization temperatures have the potential to overcome various limitations of conventional SCLCPs. Using this LC, we have designed a π-conjugated SCLCP that exhibits the N phase at room temperature.

In this study, we synthesized polyacrylate and polymethacrylate SCLCP with the **DPB[7]** mesogen, which exhibits a nematic phase at room temperature. We analyzed the detailed phase structure and evaluated the birefringence and photophysical properties. Our findings revealed several unique properties of SCLCPs that are not achievable with small molecule systems.

2. Results and Discussion

2.1. Synthesis of Monomers and Polymers

We synthesized side-chain liquid crystalline polyacrylates and polymethacrylates incorporating the 3,8-diphenyl-6,7-dihydro-5*H*-benzo[7]annulene (**DPB[7]**) skeleton with alkyl terminals (carbon number: n = 2, 4), alkoxy spacers (carbon number: m = 2, 5), and various terminal ester groups. The chemical structures and synthetic procedures of the monomers (**M1–M4**), polymers (**P1–P4**), and model compounds (**M5** and **M6**) are illustrated in Scheme 1. The synthesis of the monomers can be outlined as follows: Suzuki–Miyaura cross-coupling reaction [59,60] between 6-bromo-1-tetralone (**5**) and 4-alkylphenylboronic acid (with alkyl group carbon numbers: m = 2, 4) provided **6**, followed by a Wittig reaction to yield **7**; a ring expansion reaction using hypervalent iodine reagent [hydroxy(tosyloxy)iodo]benzene (HTIB) then produced **8**; triflation of the carbonyl group yielded **9**, and subsequent Suzuki–Miyaura cross-coupling of **9** with 4-alkoxyphenylboronic acid pinacol esters (alkoxy group carbon numbers: n = 2, 5) provided **10**; deprotection of the *tert*-butyldimethylsilyl (TBDMS) group of **10** using tetrabutylammonium fluoride (TBAF) resulted in **11**. Finally, the introduction of various acyl chlorides afforded the acrylic monomers (**M1** and **M3**), methacrylic monomers (**M2** and **M4**), and isobutyrate derivatives (**M5** and **M6**). The monomers were purified via column chromatography on silica gel and recrystallization. The chemical structures of the monomers and model compounds were confirmed through ^{1}H-NMR, ^{13}C-NMR, FT-IR, and high-resolution mass spectrometry (HRMS). These spectral data and the synthesis of **1–4** are provided in Supporting Information (SI).

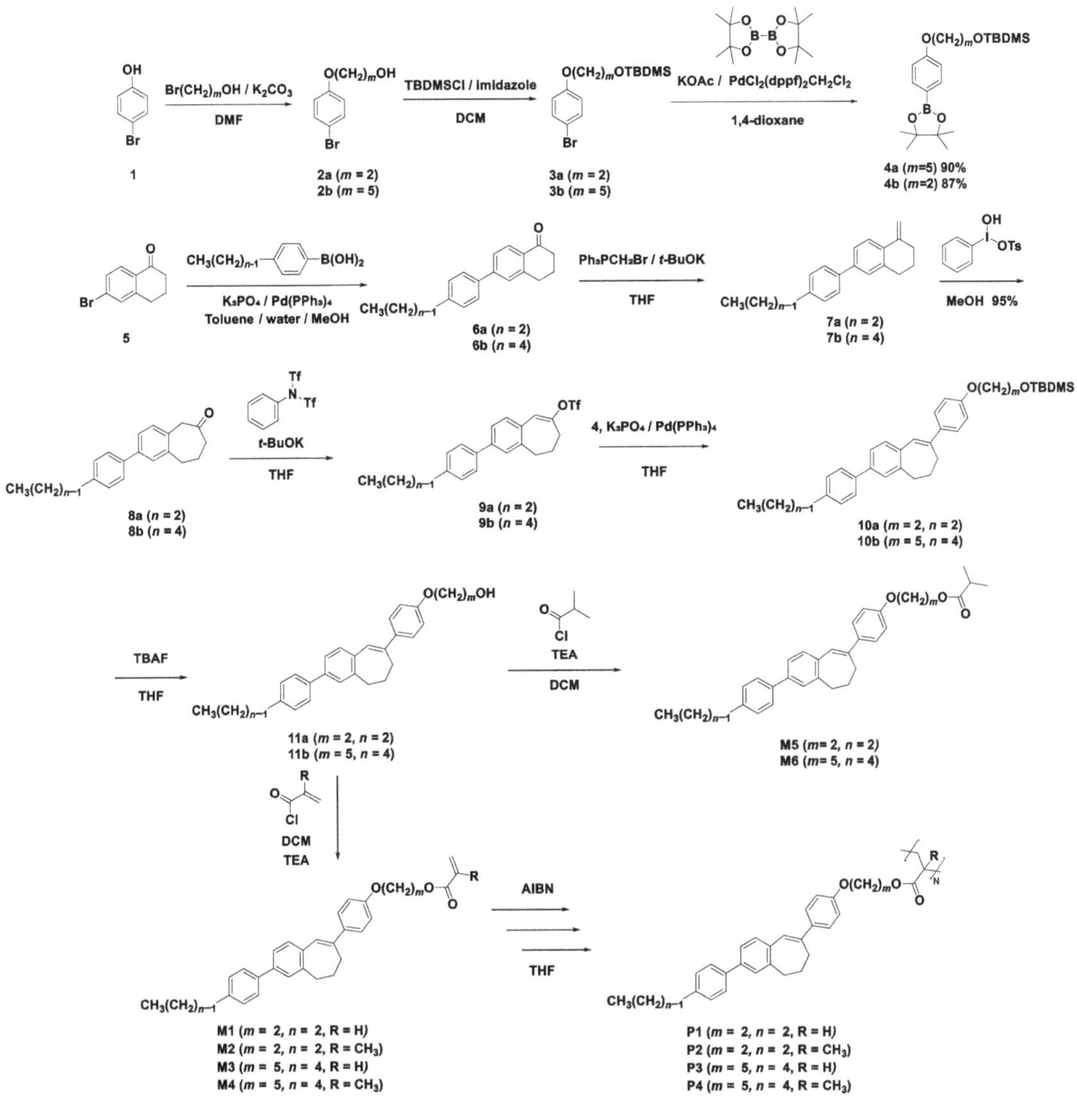

Scheme 1. Synthesis of acrylic monomers (**M1** and **M3**), methacrylic monomers (**M2** and **M4**), and model compounds of monomer (**M5** and **M6**) with **DPB[7]** skeleton and flexible chains (m + 1, n), and corresponding polymers (**P1–P4**).

Radical polymerization of the acrylic monomers (**M1** and **M3**) and methacrylic monomers (**M2** and **M4**) was conducted using 6.5 wt% azobisisobutyronitrile (AIBN) as the initiator, resulting in the corresponding polymers (**P1–P4**). The crude polymers were purified by reprecipitation in methanol. The polymer structures were confirmed by ^{1}H-NMR and ^{13}C-NMR spectroscopy. The number-average molecular weight (M_n) and weight-average molecular weight (M_w) of the polymers were determined via gel permeation chromatography using polystyrene standards. Furthermore, thermogravimetric analysis (TGA) measurements were performed on **P1–P4** at a rate of 20 °C min^{-1} (Figures S1–S4), determining degradation temperatures at a 10% weight loss (T_{10}). The results are summarized in Table 1. The polymethacrylates exhibited higher molecular weights compared with the polyacrylates.

Table 1. The results of polymerization: yields, M_n, M_w, polydispersity indexes (M_w/M_n), and degradation temperature at 10% weight loss (T_{10}) of **P1–P4**.

Entry	Yield (%)	M_n	M_w	M_w/M_n	T_{10} [°C]
P1	40	3000	5200	1.70	410
P2	34	10,400	20,000	1.93	409
P3	19	5000	7000	1.40	413
P4	10	17,100	26,500	1.55	412

2.2. Phase Transition Behaviors

Liquid crystalline phases of the polymers, monomers, and model compounds were identified using polarized optical microscopy (POM). Phase transition temperatures and enthalpies were determined through differential scanning calorimetry (DSC). POM observations suggested that **M1** and **M2** exhibited only the nematic (N) phase, while **M3** and **M4** exhibited both the N phase and the smectic A (SmA) phase. It is suspected that the vinyl groups in **M1–M4** underwent either conversion to another compound or oligomerization owing to thermal reactions. Therefore, we synthesized model compounds (**M5** and **M6**) in which the (meth)acrylate group was replaced with an isobutyrate group. **M5** and **M6** were used for comparison with the monomers and polymers in subsequent analyses. The phase transition temperatures and enthalpies (ΔH) of **M5** and **M6** and **P1–P4** are summarized in Table 2, and the typical POM images of **P2** and **P4** are illustrated in Figure 1. The DSC curves for **M5** (second scan), **M6** (second scan), and **P1–P4** (third scan) are shown in Figures S5–S10, while the POM images for **M5**, **M6**, **P1**, **P3**, and the rest of **P2** and **P4** are shown in Figures S11–S16.

Table 2. Phase transition temperature [°C] (enthalpy ΔH [kJ mol^{-1}]) and the range of LC phase (ΔT_{LC} [°C]) of **DPB[7]** monomers (**M5** and **M6**) and **DPB[7]** polymers (**P1–P4**) upon heating and cooling at a rate of 10 °C min^{-1} (Cr and Cr2: crystal phase, N: nematic phase, SmA: smectic A phase, Iso: isotropic phase, G: glass state).

Entry	Phase Transition Behavior	
	Heating [a]	Cooling [b]
DPB[7]-C3 [c]	Cr 52.6 (11.8) N 160.4 (0.85) Iso	Iso 158.3 (0.86) N 4.8 (5.68) Cr
M5	Cr 37.9 (0.15) Cr2 [d] 66.4 (0.24) N 111.6 (0.33) Iso	Iso 109.8 (0.60) N
M6	Cr 77.4 (22.5) SmA 84.8 (0.49) N 109.3 (0.77) Iso	Iso 107.5 (0.41) N 83.0 (0.49) SmA 29.5 (15.6) Cr
P1	G 60.6 N 176.3 (0.22) Iso	Iso 173.4 (0.36) N 57.5 G
P2	N 205.0 (0.32) Iso	Iso 202.5 (0.28) N
P3	G 25.6 SmA 150.3 (0.65) N 184.2 (0.32) Iso	Iso 181.8 (0.46) N 152.8 (1.28) SmA 23.2 G
P4	SmA 193.8 (0.75) N 205.0 (0.19) Iso	Iso 202.0 (0.44) N 181.5 (1.18) SmA

[a] Third heating for **DPB[7]-C3** and **P1–P4**, and second heating for **M5** and **M6**. [b] Third cooling for DPB[7]-C3 and **P1–P4** and second cooling for **M5** and **M6**. [c] Reported by Iwai et al. [54]. [d] A crystal–crystal phase transition was observed only for **M5**.

M5 exhibited the N phase between 66.4 °C and 111.6 °C during the heating process, while **M6** exhibited the SmA phase between 77.4 °C and 84.8 °C and the N phase between 84.8 °C and 109.3 °C. POM images of the schlieren textures in the N phase for **M5** and **M6** are illustrated in Figures S11 and S12, and fan-shaped textures in the SmA phase of **M6** are also presented in Figure S12. The flexible chain length, excluding the terminal ester group, (in terms of carbon and oxygen atoms) is three or fewer for **M5** and four or more for **M6**. This trend is similar to our **DPB**[7] liquid crystals [58], where the SmA phase emerges when the flexible chain length exceeds three atoms.

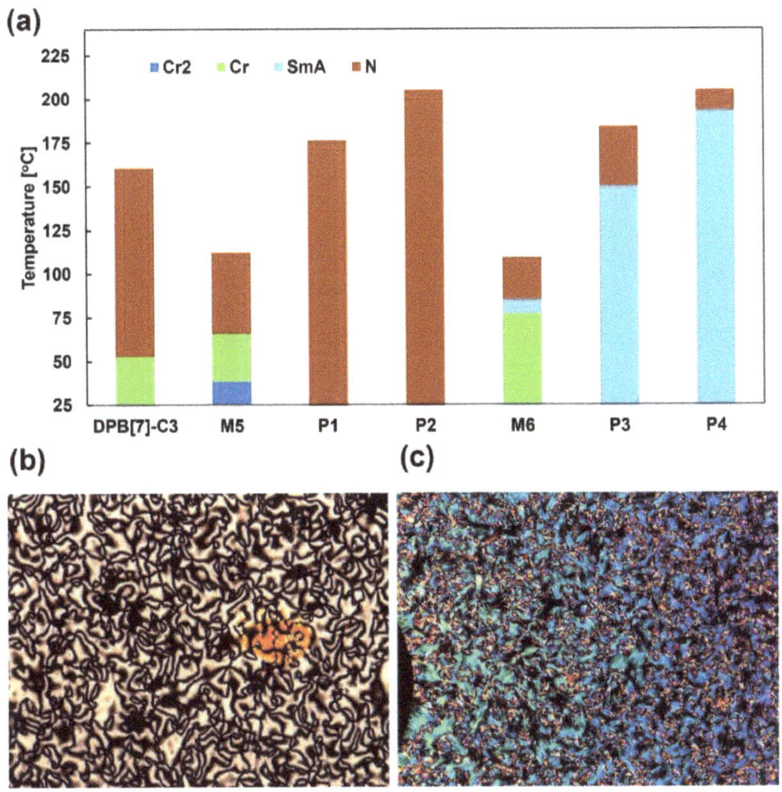

Figure 1. Summary of phase transition behavior. (**a**) Phase transition temperature and range diagram for **DPB[7]** monomers (**DPB[7]-C3**, **M5**, and **M6**) and **DPB[7]** polymers (**P1–P4**) observed by DSC measurement in second heating (for **M5** and **M6**) and third heating (for **DPB[7]-C3** and **P1–P4**) process. POM images of (**b**) schlieren texture for **P2** at 193 °C and (**c**) fan-shape texture for **P4** at 128 °C.

Next, we describe the phase transition behaviors of **P1–P4**. **P1** and **P2** exhibited N phases over the ranges of 25.0–176.3 °C and 25.0–205.0 °C, respectively, during the heating process. In these polymers, the N phase was successfully maintained at room temperature (25 °C), and schlieren textures were observed, as illustrated in Figure 1b. **P3** displayed a smectic A (SmA) phase between 25.0 °C and 150.3 °C and an N phase from 150.3 °C to 184.2 °C, while **P4** showed a SmA phase from 25.0 °C to 193.8 °C and an N phase from 193.8 °C to 205.0 °C. In the SmA phases of **P3** and **P4**, fan-shaped textures were observed (Figure 1c). The glass transition temperature (T_g) was observed at 60.6 °C for P1, and 25.6 °C for **P3**, while no T_g was detected for **P2** and **P4**.

The clearing points of **P1** and **P2** were 64.7 °C and 93.4 °C higher than those of **M5**, respectively. Similarly, the clearing points of **P3** and **P4** were higher than those of **M6** by 74.9 °C and 95.7 °C, respectively, and the N-SmA phase transition temperatures were higher by 65.5 °C and 109.0 °C. These results, indicating higher phase transition temperatures due to the polymer effect, are consistent with previously reported findings for SCLCPs. Notably, these liquid crystalline phases were maintained even after more than three months, as illustrated in Figures S13 and S14. The room temperature N phase observed in **P1** and **P2** offers significant advantages for optoelectronic applications operating at room temperature. Finally, upon 25.0 °C, the LC temperature ranges (ΔT_{LC}) for **P1–P4** were 151.3 °C, 180.0 °C,

159.2 °C, and 180.0 °C, respectively. These values were considerably larger than those of the model monomers **M5** and **M6** (ΔT_{LC} = 45.2 °C and 31.9 °C, respectively).

2.3. Structure Analysis of LC Phase

To investigate changes in the molecular arrangement of the LCs due to polymer effect, we performed wide-angle X-ray diffraction (WAXD) measurements for **P1–P4** and the corresponding monomers **M5** and **M6** (Figures S17–S34). The WAXD profiles are summarized in Figure 2, in which the magnetic field is in the vertical direction of the paper. In the N phase, the peak observed in the small-angle region of the WAXD profile ($2\theta_{small}$) corresponded to the layer spacing (d_{LC}) along the molecular long axis. Also, in the SmA phase, not only the first-order peak ($2\theta_{small}$) but also second- or third-order peaks were observed for P3 and P4, while the peak in the wide-angle region ($2\theta_{wide}$) corresponded to the average distance between mesogens ($d_{mesogen}$) along the molecular short axis. The values of the small-angle and wide-angle peaks ($2\theta_{small}$, $2\theta_{wide}$), along with the corresponding *d*-spacing values (calculated using Bragg's law) and the molecular length/side-chain length (*l*) (obtained from DFT calculations (Figures S35 and S36, and Tables S1 and S2)) are presented in Table 3.

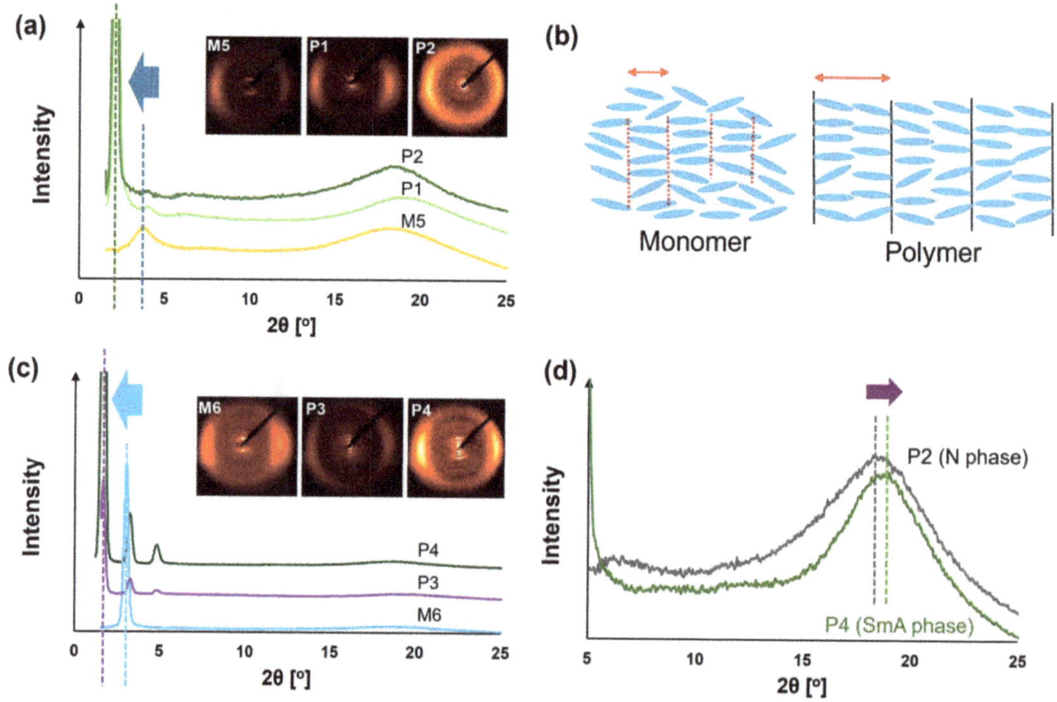

Figure 2. WAXD intensity profiles and expected structure in LC phase. (**a**) WAXD intensity profile in nematic (N) phase for **M5** (80 °C), **P1** (40 °C), and **P2** (30 °C). The $2\theta_{small}$ values are shifted to small angle region due to the polymer effect. (**b**) Schematic illustration of molecular packing in N phase with cybotactic cluster for monomer (monolayer) and polymer (double-layer). (**c**) WAXD intensity profile in smectic A phase for **M6** (80 °C), **P3** (30 °C), and **P4** (30 °C) The $2\theta_{small}$ values are shifted to small angle region due to the polymer effect. (**d**) WAXD intensity profile for **P2** and **P4** in wide-angle region recorded at 30 °C. The $2\theta_{wide}$ value for SmA phase is slightly larger than that for N phase.

Table 3. The 2θ peak angle in small-angle region ($2\theta_{small}$ [°]) and wide-angle region ($2\theta_{wide}$ [°]) obtained from WAXD profile, their 2θ *d*-spacing values (d_{LC} and $d_{mesogen}$ [Å]), calculated molecular length (l [Å]), and the value of d_{LC}/l.

Entry	$2\theta_{small}$ [°]	d_{LC} [a] [Å]	l [b] [Å]	d_{LC}/l [-]	$2\theta_{wide}$ [°]	$d_{mesogen}$ [Å]
M5	3.60 [c]	24.6	24.2	1.0	18.5 [c]	4.78
P1	2.09	42.4	22.4	1.9	18.9 [d]	4.71
P2	2.01 [e]	44.0	22.4	2.0	18.3 [e]	4.85
M6	2.97 [c]	29.8	30.2	1.0	18.5 [c]	4.78
P3	1.61 [e]	55.1	28.9	1.9	19.3 [e]	4.60
P4	1.61 [e]	55.1	28.9	1.9	18.9 [e]	4.69

[a] Layer spacing of LC phase (nematic phase for **M5**, **P1**, **P2**, and smectic A phase for **M6**, **P3**, **P4**) calculated from $2\theta_{small}$ [°] value. [b] Whole molecular length (or length of side-chain for polymers) when alkyl chains are assumed to be fully extended in all-*trans* (calculated using DFT calculation at the B3LYP/6-31G(d) level of theory). [c] Measured at 80 °C. [d] Measured at 40 °C. [e] Measured at 30 °C.

In the N phase of **M5** (at 80 °C), **P1** (at 40 °C), and **P2** (at 30 °C), broad scattering peaks in the wide-angle region and sharp peaks in the small-angle region were observed (Figure 2a). In the N phase, there was no long-range positional order, and typically, only broad scattering in the wide-angle region was detected in WAXD measurements. Therefore, the observed small-angle peaks suggest the formation of smectic-like micro aggregates, specifically cybotactic clusters. The azimuthal profiles of diffraction in the small-angle region for **M5**, **P1**, and **P2** (Figures S19, S25 and S28) demonstrated that the scattering in the small-angle region was orthogonal to that in the wide-angle region. These results indicate that the cybotactic clusters [61–63] are SmA-type, with no uniform tilt in the domain. Notably, the ratio of the small-angle peak intensity to the wide-angle peak intensity was more than 10 times greater in **P1** and **P2** than in **M5**, suggesting that cybotactic clusters form more readily in **P1** and **P2** than in **M5**.

The $2\theta_{small}$ values for **M5**, **P1**, and **P2** were 3.60°, 2.09°, and 2.01°, respectively, corresponding to d_{LC} values of 24.6 Å, 42.4 Å, and 44.0 Å, respectively. DFT calculations revealed that the molecular lengths (l) for **M5**, **P1**, and **P3** were 24.2 Å, 22.4 Å, and 22.4 Å, respectively, when the flexible chains were in the all-*trans* configuration and fully extended. Consequently, the d_{LC}/l ratios for **M5**, **P1**, and **P3** were calculated to be 1.0, 1.9, and 2.0, respectively. This indicates that the cybotactic clusters in the N phase of **M5** form a monolayer structure, while those in **P1** and **P3** form a double-layer structure (Figure 2b).

In the SmA phase of **M6** (80 °C), **P3** (30 °C), and **P4** (30 °C), broad scattering peaks were observed in the wide-angle region, while sharp peaks appeared in the small-angle region. The azimuthal profiles of diffraction in the small-angle region for **M6**, **P3**, and **P4** (Figures S22, S31, and S34) showed that the scattering in the small-angle region was orthogonal to that in the wide-angle region, a characteristic of the SmA phase. The $2\theta_{small}$ values for **M6**, **P3**, and **P4** were 2.97°, 1.61°, and 1.61°, respectively, corresponding to d_{LC} values of 29.8 Å, 55.1 Å, and 55.1 Å. DFT calculations revealed l = 30.2 Å, 28.9 Å, and 28.9 Å for **M6**, **P3**, and **P4**, respectively, from which the d_{LC}/l ratios were determined to be 1.0, 1.9, and 1.9. This suggests that **M6** has a monolayer structure in the SmA phase, while **P3** and **P4** exhibit double-layer structures (Figure 2b).

In summary, **M5** and **M6** exhibit monolayer structures in the LC phase, while **P1–P4** form double-layer structures. The formation of a double-layer structure is commonly reported in SCLCPs [64,65]. The $2\theta_{wide}$ values observed in the N phase of **M5**, **P1**, and **P2** were 18.5°, 18.9°, and 19.3°, respectively, while those observed in the SmA phase of **M6**, **P3**, and **P4** were 18.5°, 18.9°, and 19.3°, respectively. Consequently, $d_{mesogen}$ values for **M5**, **P1**, **P2**, **M6**, **P3**, and **P4** were calculated to be 4.78 Å, 4.71 Å, 4.85 Å, 4.78 Å, 4.60 Å, and 4.69 Å, respectively. Although there was no difference between the $d_{mesogen}$ values of **M5** and **M6**, the values of **P3** and **P4** were 0.02–0.25 Å smaller than those of **P1** and **P2**.

2.4. Birefringence Properties

To evaluate the optical properties of **P1–P4**, temperature-variable birefringence (Δn) measurements were performed following our previously reported method [66–70]. For comparison with the corresponding polymers, Δn measurements were also conducted for **M5** and **M6** because of the lack of heat stability of **M1–M4**. First, the nematic liquid crystals (NLCs) were filled into homogeneously aligned polyimide cells. When preparing LC cells for **P1–P4**, the acrylic polymers **P1** and **P3** could be filled into the polyimide cells via capillary force; however, the methacrylic polymers **P2** and **P4** could not be filled due to their high viscosity, likely caused by differences in molecular weight. To confirm uniaxial orientation, POM observations were carried out. The prepared LC cells appeared dark when the polarizer was aligned with the rubbing direction and brightest when at 45° to the rubbing direction (Figure S37). These results confirm that the nematic director was aligned with the rubbing direction.

Next, we employed a micro-spectroscopic method, observing the transmitted light through the LC cell as a function of wavelength (λ) under cross-polarized conditions. The nematic director was set at 45° to the polarizer. A typical transmission light plot is shown in Figure S38 (dot). The transmitted light intensity (I) was fitted using the following Equation (1) and Cauchy's Equation (2) to determine the coefficients a, b, and c, where A is a constant.

$$\frac{I}{I_0} = A\sin^2\left(\frac{\pi \Delta n d}{\lambda}\right)\sin^2\theta \tag{1}$$

$$\Delta n = a + \frac{b}{\lambda^2} + \frac{c}{\lambda^4} \tag{2}$$

The theoretical curve obtained from Equation (2) fitted well to the transmitted light plot (Figure S38 (solid line)). From this fitting, we obtained the Δn data presented in Figure 3. Figure 3a illustrates the wavelength dependence of Δn at T/T_i = 0.9 (T_i; isotropic temperature), while Figure 3b shows Δn plotted against temperature for **M5**, **M6**, **P1**, and **P3** at 550 nm. As shown in Figure 3a, the Δn values of **P1** and **P3** were higher than those of **M5** and **M6**, indicating that Δn increased as a result of polymerization. Additionally, **P1** and **P3** exhibited Δn values of 0.24 and 0.20 at room temperature (25 °C), respectively.

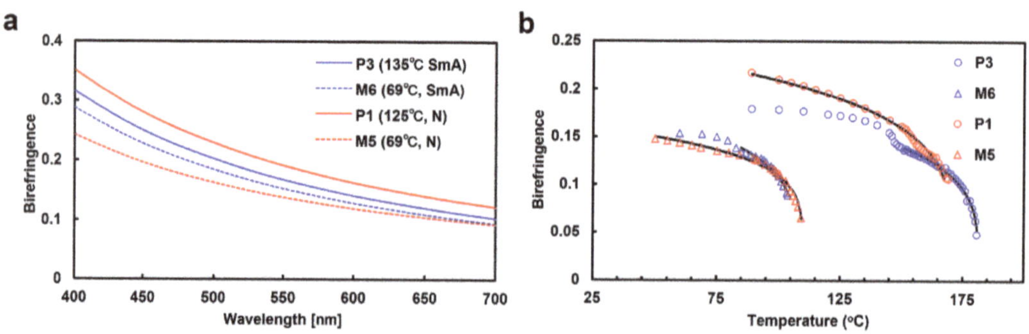

Figure 3. Birefringence is dependent on (**a**) wavelength and (**b**) temperature for **M5**, **M6**, **P1**, and **P3**. (**a**) Measured at the same reduced temperature (T/T_i = 0.9) in each LC phase. (**b**) Plotted against temperature at 550 nm and fitted by Equations (3) and (4). For **M6** and **P3**, the fitting was performed within the nematic phase.

As illustrated in Figure 3b, Δn decreased with increasing temperature. The relationship between Δn and temperature can be explained by the temperature dependence of the order parameter (S) of the nematic director, which is described by Haller's approximation:

$$\Delta n = \Delta n_0 S \tag{3}$$

$$S = \left(1 - \frac{T}{T_i}\right)^{\beta} \quad (4)$$

where Δn_0 is the extrapolated value for the perfectly oriented birefringence (S = 1) of the N LC, T_i is the clearing point, and β is a material constant characteristic of the NLC. The theoretical curve obtained from Equation (4) fitted well to the plot of Δn for each compound (Figure 3), and the values of β and Δn_0 were determined. Additionally, the order parameter S was calculated using Equation (3) from the measured Δn values and the corresponding Δn_0 values.

The values of S at 90 °C were calculated, as this was the temperature at which **M5**, **M6**, **P1**, and **P3** could be measured, and are shown in Table 4 along with the corresponding values of β and Δn_0. The Δn_0 values for **M5**, **M6**, **P1**, and **P3** were 0.21, 0.26, 0.31, and 0.26, respectively, and the β values were 0.18, 0.23, 0.22, and 0.23, respectively. The S values at 90 °C were 0.60, 0.50, 0.69, and 0.69, respectively. Comparison of these values indicates an improvement in S due to the polymer effect [21].

Table 4. Determined values of Δn_0 and β by Haller's approximation and estimated birefringence (Δn) and order parameter (S) at 90 °C.

Entry	Δn_0	β	Δn (90 °C)	S (90 °C)
M5	0.21	0.18	0.12	0.60
M6	0.26	0.23	0.13 [a]	0.50 [b]
P1	0.31	0.22	0.22	0.69
P3	0.26	0.23	0.18 [a]	0.69 [b]

[a] Value in smectic A phase. [b] calculated using Δn (90 °C) value in smectic A phase.

Furthermore, a comprehensive evaluation of the S values, combined with the results from DSC and WAXD measurements, led to the conclusion that the significant increase in the higher phase transition temperatures (SmA–N and N–isotropic transitions) of **P1–P4** might be attributed to the enhancement of S, which resulted from the formation of a double-layer structure in the liquid crystalline phase.

2.5. Fluorescence Properties

To investigate the fluorescence properties of **M1–M4** and **P1–P4**, absorption spectra, fluorescence spectra, and fluorescence quantum yields (Φ_{fl}) were measured in a dilute THF solution and in the solid state. Additionally, the fluorescence spectra of cast films of **P1–P4** were measured. The obtained values for maximum absorption wavelength (λ_{abs}), maximum fluorescence wavelength (λ_{fl}), and Φ_{fl} for **M1–M4** and **P1–P4** are summarized in Table 5.

The λ_{abs} values for all the monomers (**M1–M4**) in THF were 316 nm. The Φ_{fl} values for **M1–M4** in THF solution were 0.02, while in the solid state, they were 0.43, 0.58, 0.61, and 0.48, respectively. This indicates that **M1–M4** exhibited aggregation-induced emission (AIE) properties. The AIE properties were further confirmed by aggregation experiments on **M2** and **P2** (Figure 4). The λ_{fl} values for **M1–M4** in the THF solution were approximately 400 nm, while in the solid state (polycrystalline state), they were 417, 407, 413, and 408 nm, respectively. These results suggest that the fluorescence in the solid state may be derived from aggregates rather than isolated monomers. This behavior was similar to that of our previously reported AIEgen, bridged stilbene [55].

The λ_{abs} value for **P1–P4** in THF was 316 nm. The Φ_{fl} value for **P1–P4** in the THF solution was approximately 0.03, while in the solid state, the values were 0.14, 0.15, 0.35, and 0.18, respectively. **P1–P4** exhibited the same AIE properties as the monomers, and aggregation experiments with **P2** and **P4** yielded similar results to those of the monomers (Table S3). Similar redshifts in the λ_{fl} values for **P1–P4** in both the THF solution and the solid state were observed, mirroring the behavior of the monomers. These findings revealed that the luminescence properties of the bulk polymers and monomers were nearly identical. Interestingly, the λ_{fl} values for **P1–P4** in the polymer films were 442, 456, 469, and 473 nm, respectively, redshifted by more than 27 nm compared with those in the solid state (polycrystalline state) (Figure 5). The relationship between the λ_{fl} of the films and the liquid crystalline phase at room temperature was also investigated. The λ_{fl} values for **P1** and **P2** (which exhibited the N phase) were around 450 nm, while those for **P3** and **P4** (which exhibited the SmA phase) were around 470 nm. These results suggest that the λ_{fl} in the films may vary depending on the liquid crystalline phase, with a relatively shorter wavelength shift in the N phase and a longer wavelength shift in the SmA phase. This indicates that the molecular orientation and interactions in different LC phases may affect the fluorescence properties.

Table 5. Spectroscopic properties for **M1–M4** and **P1–P4** in THF solution, solid state, and film.

Entry	λ_{abs} [nm]	Φ_{fl} [-]		λ_{fl} [nm]		
	THF	THF	Solid [a]	THF	Solid [a]	Film
M1	316	0.02	0.43	399	417	-
M2	316	0.02	0.58	399	407	-
M3	316	0.02	0.61	401	413	-
M4	316	0.02	0.48	400	408	-
P1	316	0.03	0.14	407	415	442
P2	316	0.02	0.16	399	413	456
P3	316	0.03	0.35	400	413	469
P4	316	0.03	0.18	407	413	473

[a] Polycrystalline solid.

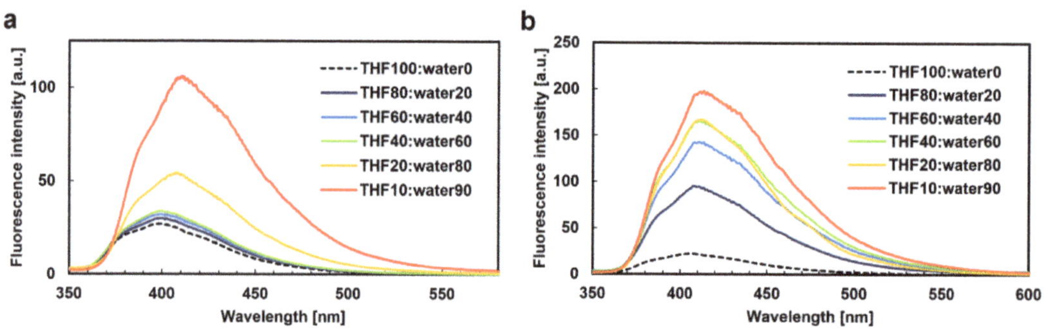

Figure 4. Aggregation experiments. Fluorescence spectra of (**a**) **M2** and (**b**) **P2** in THF/water mixtures with different water contents (vol %) excited at each λ_{abs}; the concentration is 10^{-5} M.

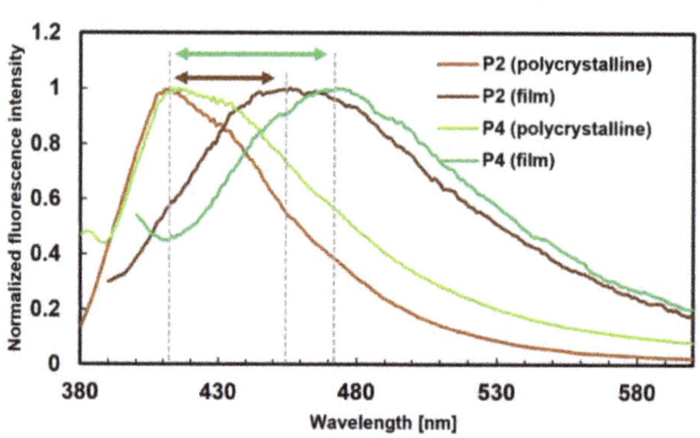

Figure 5. Fluorescence spectra of **P2** and **P4** excited at 370 nm (polycrystalline) and at 380 nm (film).

3. Experimental

3.1. Materials

Unless otherwise noted, all solvents and chemicals were commercially available and used without further purification. Column chromatography was performed on silica gel (Silica Gel 60N, 63–210 μm, Kanto chemical Co., Inc., Tokyo, Japan). [Hydroxy(tosyloxyl)iodo]benzene, tetrakis(triphenylphoshine)palladium (0) ($Pd(PPh_3)_4$), [1,1-bis(diphenylphosphino)ferrocene]palladium (0) ($Pd(dppf)Cl_2 \cdot CH_2Cl_2$), and 6-bromo-1-tetralone were purchased from TCI (Tokyo, Japan).

3.2. Instruments

^{1}H-NMR and ^{13}C-NMR spectra were recorded on BRUKER 500 (Yokohama, Japan) (500 MHz) and JEOL 400 (Tokyo, Japan) (100 MHz) spectrometers, respectively, for $CDCl_3$ solution using tetramethylsilane (TMS) as an internal standard. ^{1}H-NMR spectra were reported as follows: chemical shift (δ ppm), multiplicity (s = singlet, d = doublet, t = triplet, q = quartet, m = multiplet), integration, and coupling constants in units of Hz. ^{13}C-NMR spectra were reported as chemical shifts in ppm. The FT-IR spectra were recorded on a JASCO FT-IR 4600 spectrometer (Tokyo, Japan). High-resolution EI mass spectra (HRMS) were recorded on a double-focusing mass spectrometer JEOL JMS-700, measured at the Tokyo Institute of Technology Open Facility Center. This center is independent of our laboratory to ensure fairness. Size exclusion chromatography (SEC) was performed using a JASCO system (PU-2080, CL-NETII/ADC, CO-2060, UV-2075, RI-2031, Tokyo, Japan) equipped with TSK gel columns (TOSOH G3000H xl, Tokyo, Japan) with THF as the eluent at the following rate of 0.85 mL min^{-1} at 40 °C after calibration with polystyrene standards. Polarized optical microscopy (POM) was performed using a Leica DM2500P microscopy (Wetzlar, Germany) with a Mettler FP90 hot stage (Greifensee, Switzerland). Differential scanning calorimetry (DSC) was performed using PerkinElmer DSC 8500 equipment (Waltham, MA, USA) at a scanning rate of 10 °C min^{-1} under a flow of dry nitrogen. Thermo-gravimetric analysis (TGA) was performed using a Rigaku Thermo Plus EVO2 series TG-DTA 8122 (Tokyo, Japan) at a heating rate of 20 °C min^{-1} under a flow of dry nitrogen. The initial mass of the samples was 3–6 mg. X-ray investigations were carried out with samples kept in glass capillary tubes (1.0 mm diameter) for oriented patterns under a magnetic field. Wide-angle X-ray diffraction (WAXD) patterns were obtained using a Bruker D8 DISCOVER (Billerica, MA, USA) equipped with a Vantec-500 detector and Cu Kα radiation. UV-Vis spectra were recorded on a JASCO V-670 UV-vis spectrophotometer (Tokyo, Japan) and fluorescence spectra were recorded on a JASCO FP-6500 spectrofluorometer (Tokyo, Japan). Absolute quantum yields were measured by Hamamatsu Photonics Quantaurus QY apparatus (Hamamatsu City, Japan). All sample solutions were de-aerated by bubbling with argon for 15 min prior to the quantum yield measurement.

3.3. Birefringence

Measurement of birefringence was performed in uniaxially aligned LC cells containing indium tin oxide (ITO) purchased from EHC. The cell gap (*d*) of 3–5 μm was determined by the interferometric method [66–70].

3.4. Theoretical Calculations

Theoretical calculations were carried out on the Gaussian 16 program [71]. Geometry optimizations were carried out using DFT methods at the B3LYP with the 6-31G(d) basis. Whether the optimized geometry was at the stationary point without any imaginary frequency was checked by the frequency calculation performed at the optimized geometries using their level of theory.

3.5. Synthesis

6-(4-ethylphenyl)-3,4-dihydronaphthalen-1(2*H*)-one (**6a**)

To a solution of 6-bromo-1-tetralone (**5**) (3.4 g, 15 mmol), 4-alkylphenylboronic acid (3.4 g, 22 mmol), potassium phosphate (9.5 g, 45 mmol) in solvent (30 mL/12 mL/6 mL; toluene/water/methanol), $Pd(PPh_3)_4$ (0.55 g, 0.48 mmol) was added under argon atmosphere, and the mixture was refluxed (100 °C) for 3 hours. After the reaction, the mixture was cooled to room temperature, then extracted with dichloromethane. The organic layer was washed with water three times, dried over $MgSO_4$, filtrated, and evaporated in vacuo. The residue was purified by column chromatography on silica gel (6/1 (*v*/*v*) hexane/ethyl acetate) to afford **6a** as a brown solid; yield 96%; ^{1}H-NMR (500 MHz, $CDCl_3$) δ 8.09 (d, *J* = 8.2 Hz, Ar-*H*, 1H), 7.56–7.52 (m, Ar-*H*, 3H), 7.47–7.45 (m, Ar-*H*, 1H), 7.31–7.29 (m, Ar-*H*, 2H), 3.02 (t, *J* = 6.1 Hz, -CH_2-, 2H), 2.73–2.67 (m, -CH_2-, 4H), 2.20–2.15 (m, -CH_2-, 2H), 1.28 (t, *J* = 7.6 Hz, -CH_3, 3H) ppm (Figure S47).

6-(4-ethylphenyl)-1-methylene-1,2,3,4-tetrahydronaphthalene (**7a**)

Methyltriphenylphosphonium bromide (7.1 g, 19 mmol) was dissolved in THF (40 mL) under argon atmosphere. The solution was cooled to 0 °C, and then potassium tert-butoxide (2.4 g, 21 mmol) was added. Following stirring for 10 minutes, **6a** (14.5 mmol) was added, warming to room temperature and stirring overnight. The reaction mixture was quenched by adding NH_4Cl aq to the solution and extracted with ethyl acetate. The organic layer was washed with water three times, dried over $MgSO_4$, filtrated, and evaporated in vacuo. The residue was purified by column chromatography on silica gel (4/1 (*v*/*v*) hexane/dichloromethane) to afford **7a** as a colorless solid; yield 90%; ^{1}H-NMR (500 MHz, $CDCl_3$) δ 7.71 (d, *J* = 8.2 Hz, Ar-*H*, 1H), 7.51 (d, *J* = 8.2 Hz, Ar-*H*, 2H), 7.40–7.37 (m, Ar-*H*, 1H), 7.33–7.32 (m, Ar-*H*, 1H), 7.26 (d, *J* = 7.6 Hz, Ar-*H*, 2H), 5.51 (s, C=C*H*, 1H), 4.96 (s, C=C*H*, 1H), 2.90 (t, *J* = 6.3 Hz, -CH_2-, 2H), 2.69 (q, *J* = 7.6 Hz, -CH_2-, 2H), 2.57 (t, *J* = 6.1 Hz, -CH_2-, 2H), 1.94–1.89 (m, -CH_2-, 2H), 1.27 (t, *J* = 7.6 Hz, -CH_3, 3H) ppm (Figure S49).

2-(4-ethylphenyl)-5,7,8,9-tetrahydro-6*H*-benzo[7]annulen-6-one (**8a**)

To a solution of **7a** (13 mmol) in solvent (38 mL/2 mL; methanol/water), HTIB ([hydroxy(tosyloxy)iodo]benzene; 5.6 g, 14 mmol) was added under air, and stirred at room temperature for 20 min. After the reaction, the mixture was extracted with dichloromethane. The organic layer was washed with water three times, dried over $MgSO_4$, filtrated, and evaporated in vacuo. The residue was purified by column chromatography on silica gel (6/1 (*v*/*v*) hexane/ethyl acetate) to afford **8a** as a colorless solid; yield 92%; ^{1}H-NMR (500 MHz, $CDCl_3$) δ 7.51 (d, *J* = 8.2 Hz, Ar-*H*, 2H), 7.41–7.38 (m, Ar-*H*, 2H), 7.27 (d, *J* = 8.2 Hz, Ar-*H*, 2H), 7.21 (d, *J* = 7.6 Hz, Ar-*H*, 1H), 3.76 (s, -CH_2-, 2H), 3.01 (t, *J* = 6.4 Hz, -CH_2-, 2H), 2.69 (q, *J* = 7.6 Hz, -CH_2-, 2H), 2.60 (t, *J* = 7.0 Hz, -CH_2-, 2H), 2.06–2.01 (m, -CH_2-, 2H), 1.28 (t, *J* = 7.6 Hz, -CH_3, 3H) ppm (Figure S51).

3-(4-ethylphenyl)-6,7-dihydro-5*H*-benzo[7]annulen-8-yl trifluoromethanesulfonate (**9a**)

The mixture of **8a** (12 mmol) and THF (30 mL) was cooled to −20 °C, and then tert-butoxide (2.4 g, 21 mmol) was added. The mixture was stirred at 0 °C for 1 h, and then cooled to −20 °C, then N-phenylbis(trifluolomethanesulfonimide) (5.3 g, 15 mmol) was added, and stirred at −20 °C for a further 1 h, and then warmed to 0 °C, and stirred for 4 h. After that, the reaction was quenched by the dropwise addition of water (20 mL), and organic products were extracted with ethyl acetate. The organic layer was washed with water three times, dried over $MgSO_4$, filtrated, and evaporated in vacuo. The residue was purified by column chromatography on silica gel (6/1 (*v*/*v*) hexane/ethyl acetate) to afford **9a** as a slight yellow solid; yield 98%; ^{1}H-NMR (500 MHz, $CDCl_3$) δ 7.51 (d, *J* = 7.9 Hz, Ar-*H*, 2H), 7.43–7.40 (m, Ar-*H*, 1H), 7.34–7.33 (m, Ar-*H*, 1H), 7.27 (d, *J* = 8.2 Hz, Ar-*H*, 2H), 7.22 (d, *J* = 7.9 Hz, Ar-*H*, 1H), 6.62 (s, Ar-C*H*-, 1H), 2.95 (t, *J* = 5.2 Hz, -CH_2-, 2H), 2.81 (t, *J* = 6.4 Hz, -CH_2-, 2H), 2.70 (q, *J* = 7.6 Hz, -CH_2-, 2H), 2.06–2.01 (m, -CH_2-, 2H), 1.28 (t, *J* = 7.6 Hz, -CH_3, 3H) ppm (Figure S53).

tert-butyl(2-(4-(3-(4-ethylphenyl)-6,7-dihydro-5*H*-benzo[7]annulen-8-yl)phenoxy)ethoxy)-dimethylsilane (**10a**)

To a solution of 9 (8 mmol), 4a (16 mmol), $K_3PO_4 \cdot nH_2O$ (2.4 g, 16 mmol) in THF (40 mL), $Pd(PPh_3)_4$ (0.28 g, 0.24 mmol) was added under argon atmosphere, and stirred at 50 °C overnight. After the reaction, the mixture was cooled to room temperature, then extracted with dichloromethane. The organic layer was washed with water three times, dried over $MgSO_4$, filtrated, and evaporated in vacuo. The residue was purified by column chromatography on silica gel (4/1 (*v*/*v*) hexane/dichloromethane) to afford **10a** as a colorless solid; yield 54%; ^{1}H-NMR (500 MHz, $CDCl_3$) δ 7.55 (d, *J* = 7.9 Hz, Ar-*H*, 2H), 7.45–7.39 (m, Ar-*H*, 4H), 7.27–7.26 (m, 3H), 6.91 (d, *J* = 8.5 Hz, Ar-*H*, 2H), 6.77 (s, Ar-*H*, 1H), 4.07 (t, *J* = 5.3 Hz, -CH_2-, 2H), 3.99 (t, *J* = 5.0 Hz, -CH_2-, 2H), 2.87 (t, *J* = 6.0 Hz, -CH_2-, 2H), 2.72–2.66 (m, -CH_2-, 4H), 2.23 (t, *J* = 6.3 Hz, -CH_2-, 2H), 1.28 (t, *J* = 7.6 Hz, -CH_3, 3H), 0.92 (s, CH_3, 9H), 0.12 (s, -CH_3, 6H) ppm (Figure S55).

2-(4-(3-(4-ethylphenyl)-6,7-dihydro-5*H*-benzo[7]annulen-8-yl)phenoxy)ethan-1-ol (**11a**)

To a solution of **10a** (2.6 mmol) in THF (5.0 mL), 12M HCl aq (4.0 mL) was added, and stirred at room temperature for 10 min. After that, the reaction was quenched by the dropwise addition of $NaHCO_3$ aq (20 mL), and organic products were extracted with dichloromethane. The organic layer was washed with water three times, dried over $MgSO_4$, filtrated, and evaporated in vacuo. Purification by recrystallization ($CHCl_3$/Hex) gave **11a** as a colorless solid; yield 87%; ^{1}H-NMR (500 MHz, $CDCl_3$) δ 7.55 (d, *J* = 8.2 Hz, Ar-*H*, 2H), 7.46 (d, *J* = 8.7 Hz, Ar-*H*, 2H), 7.42 (dd, *J* = 7.8, 2.0 Hz, Ar-*H*, 1H), 7.39 (s, Ar-*H*, 1H), 7.28–7.26 (m, 3H), 6.93 (d, *J* = 8.9 Hz, Ar-*H*, 2H), 6.78 (s, Ar-*H*, 1H), 4.12 (t, *J* = 4.6 Hz, -CH_2-, 2H), 4.00–3.97 (m, -CH_2-, 2H), 2.88 (t, *J* = 6.1 Hz, -CH_2-, 2H), 2.72–2.66 (m, -CH_2-, 4H), 2.26–2.21 (m, -CH_2-, 2H), 1.28 (t, *J* = 7.5 Hz, -CH_2-, 3H) ppm (Figure S57).

2-(4-(3-(4-ethylphenyl)-6,7-dihydro-5*H*-benzo[7]annulen-8-yl)phenoxy)ethyl acrylate (**M1**)

To a solution of compound **11a** (1.3 mmol) and triethylamine (0.27 mL, 2.0 mmol) in dichloromethane (5.0 mL) was added acyl chloride (1.7 mmol) and the mixture was stirred at room temperature for 1 h. After the reaction, the solvent was removed by evaporation, and the residue was filtered with hexane. The organic layer was washed with water three times, dried over $MgSO_4$, filtrated, and evaporated in vacuo. The residue was purified by column chromatography on silica gel (1/1 (*v*/*v*) hexane/ethyl acetate) to afford crude M1-M6. Purification by recrystallization (hexane/dichloromethane) gave pure **M1** as a colorless solid; yield: 53%; ^{1}H-NMR (500 MHz, $CDCl_3$) δ 7.55 (d, *J* = 8.2 Hz, Ar-*H*, 2H), 7.47–7.39 (m, Ar-*H*, 4H), 7.27–7.25 (m, 3H), 6.93 (d, *J* = 8.5 Hz, Ar-*H*, 2H), 6.77 (s, Ar-*H*, 1H), 6.48–6.44 (m, =*CH*, 1H), 6.18 (dd, *J* = 17.4, 10.4 Hz, -*CH*=, 1H), 5.87 (dd, *J* = 10.5, 1.4 Hz, =*CH*, 1H), 4.53 (t, *J* = 4.7 Hz, -CH_2-, 2H), 4.25 (t, *J* = 4.9 Hz, -CH_2-, 2H), 2.88 (t, *J* = 6.1 Hz, -CH_2-, 2H), 2.72–2.66 (m, -CH_2-, 4H), 2.23 (t, *J* = 6.1 Hz, -CH_2-, 2H), 1.28 (t, *J* = 7.6 Hz, -CH_3, 3H) ppm (Figure S59); ^{13}C-NMR (100 MHz, $CDCl_3$) δ 166.2, 157.9, 143.4, 142.3, 141.6, 139.2, 138.4, 137.4, 136.4, 131.5, 131.1, 128.4, 128.2, 127.7, 127.5, 127.4, 127.0, 124.5, 114.5, 66.1, 63.0, 35.0, 33.1, 30.3, 28.6, 15.7 ppm (Figure S60). HRMS (EI) Calcd for $C_{30}H_{30}O_3$: 438.5664, Found 438.2195 (Figure S57).

Poly[2-(4-(3-(4-ethylphenyl)-6,7-dihydro-5*H*-benzo[7]annulen-8-yl)phenoxy)ethyl] acrylate (**P1**)

To a pressure-resident tube, which contained **M1** (0.37 g, 0.83 mmol) and a portion of THF (3.0 mL), AIBN (azobis(isobutyronitrile), 9.0 mg, 6.5 wt%, was added, and the mixture was stirred at 60 °C for 24 h. The mixture was poured dropwise with an excess amount of methanol, and it was filtered. The solids were purified by column chromatography on Sephadex^{TM} G100 (eluted with THF), and reprecipitated with methanol; colorless solid; yield 40.3%; ^{1}H-NMR (500 MHz, $CDCl_3$) δ 7.50–7.05 (brm, Ar-H, 9H), 6.88–6.71 (brm, Ar-H, 3H), 4.26–4.10 (brm, 4H), 2.88–2.36 (brm, 6H), 2.17–2.13 (brm, 2H), 1.84 (br, 1H), 1.26 (br, 3H) ppm (Figure S71); ^{13}C-NMR (100 MHz, CDCl3) δ 174.7, 157.8, 143.3, 142.1, 141.6, 139.0,

138.3, 138.2, 137.2, 136.3, 135.9, 131.2, 128.3, 127.4, 127.3, 126.9, 125.6, 124.5, 114.5, 65.8, 62.9, 41.3, 41.2, 35.0, 34.3, 33.1, 31.6, 30.4, 30.1, 29.6, 28.6, 21.3, 15.7 ppm (Figure S72).

4. Conclusions

We synthesized **P1–P4** as SCLCPs with varying flexible chain lengths by radical polymerization of acrylate monomers containing π-extended bridged stilbene mesogens. The resulting polymers exhibited liquid crystalline phases over a temperature range exceeding 150 °C. **P1** and **P2** exhibited a nematic phase at room temperature, which was maintained for more than three months. These stable nematic phases are highly advantageous for device applications operating at room temperature. WAXD measurements revealed that the polymers possessed a double-layer structure in the liquid crystalline phase. Additionally, the polymers demonstrated larger order parameter (S) values compared with small molecules and exhibited higher birefringence. The polymers were also found to retain the AIE characteristics of the **DPB[7]** skeleton, with high quantum yields (Φ_{fl} = 0.14–0.35) in the solid state (polycrystalline). Furthermore, film formation resulted in a redshift of the maximum fluorescence wavelength (λ_{fl}) by more than 27 nm, yielding λ_{fl} values in the range of 442–473 nm. The magnitude of the redshift may be related to the liquid crystalline phase exhibited at room temperature, though further investigation is required to clarify the details. These optically and luminescent-functionalized SCLCPs hold promise for applications in luminescent semiconductors and optoelectronic materials, including holograms [72–74] and polarized light-emitting devices [75–85].

Supplementary Materials: The following supporting information can be downloaded at https://www.mdpi.com/article/10.3390/molecules29215220/s1: Figures S1–S4: TGA curves; Figures S5–S10: DSC thermogram; Figures S11–S16: POM images; Figures S17–S34: 1D- or 2D-WAXD profiles; Figures S35 and S36: Optimized structure (DFT calculation); Figure S37: POM images of nematic phase in polyimide cell; Figure S38: Wavelength dependence of light transmittance; Figure S39: Absorption spectra; Figure S40: Fluorescence spectra; Figures S41–S78: NMR spectra; Figures S79–S84: FT-IR spectra; Figures S85–S88: HRMS spectra; Tables S1 and S2: Atomic coordination and absolute energy (DFT calculation); Table S3: Aggregation experiments.

Author Contributions: Conceptualization, G.-i.K. and R.I.; methodology, M.T.; Formal analysis, Y.S. (Yoshimichi Shimomura), Y.S. (Yuki Sawatari) and M.T.; investigation, Y.S. (Yuki Sawatari), Y.S. (Yoshimichi Shimomura) and T.T.; writing—original draft, G.-i.K., Y.S. (Yuki Sawatari) and T.T.; writing—review and editing, G.-i.K.; project administration, G.-i.K. All authors have read and agreed to the published version of the manuscript.

Funding: This project was supported in part by MEXT/JSPS KAKENHI grants 23H02036 (G.-i.K.), JST PRESTO (G.-i.K.), Toshiaki Ogasawara Memorial Foundation (G.-i.K.), Iketani Science and Technology Foundation, Murata Science and Education Foundation (G.-i.K.), and Science Tokyo SPRING Foundation (Y.S.).

Institutional Review Board Statement: Not applicable.

Informed Consent Statement: Not applicable.

Data Availability Statement: The original contributions presented in this study are included in this article/Supplementary Materials, further inquiries can be directed to the corresponding author.

Acknowledgments: We thank Masato Koizumi (Materials Analysis Division, Tokyo Institute of Technology) for the HRMS measurements. This division is independent of our laboratory to ensure fairness. Riki Iwai and Y. Shimomura thank JSPS Research Fellowships for Young Scientists.

Conflicts of Interest: The authors declare no conflicts of interest.

References

1. Jackson, W.J., Jr. Liquid crystal polymers. XI. Liquid crystal aromatic polyesters: Early history and future trends. *Mol. Cryst. Liq. Cryst.* **1989**, *169*, 23. [CrossRef]
2. White, T.; Broer, D. Programmable and adaptive mechanics with liquid crystal polymer networks and elastomers. *Nat. Mater.* **2015**, *14*, 1087. [CrossRef] [PubMed]
3. Imrie, C.T.; Henderson, P.A. Liquid crystal dimers and higher oligomers: Between monomers and polymers. *Chem. Soc. Rev.* **2007**, *36*, 2096. [CrossRef] [PubMed]
4. Watanabe, J.; Hayashi, M.; Nakata, Y.; Niori, T.; Tokita, M. Smectic liquid crystals in main-chain polymers. *Prog. Polym. Sci.* **1997**, *22*, 1053. [CrossRef]
5. Watanabe, Y.; Kato, R.; Fukushima, K.; Kato, T. Degradable and nanosegregated elastomers with multiblock sequences of biobased aromatic mesogens and biofunctional aliphatic oligocarbonates. *Macromolecules* **2022**, *55*, 10285. [CrossRef]
6. Watanabe, J.; Hayashi, M. Thermotropic liquid crystals of polyesters having a mesogenic p,p′-bibenzoate unit. 1. Smectic A mesophase properties of polyesters composed of p,p′-bibenzoic acid and alkylene glycols. *Macromolecules* **1988**, *21*, 278. [CrossRef]
7. Watanabe, J.; Hayashi, M. Thermotropic liquid crystals of polyesters having a mesogenic p,p′-bibenzoate unit. 2. X-ray study on smectic mesophase structures of BB-5 and BB-6. *Macromolecules* **1989**, *22*, 4083. [CrossRef]
8. Zhang, Y.; Wang, X.; Yang, W.; Yan, H.; Zhang, X.; Han, D.; He, Y.; Li, C.; Sun, L. Programmable Complex Shape Changing of Polysiloxane Main-Chain Liquid Crystalline Elastomers. *Molecules* **2023**, *28*, 4858. [CrossRef]
9. Shaya, J.; Ribierre, J.-C.; Correia, G.; Dappe, Y.J.; Mathevet, F.; Mager, L.; Heinrich, B.; Méry, S. Control of the Organization of 4,4′-bis(carbazole)-1,1′-biphenyl (CBP) Molecular Materials through Siloxane Functionalization. *Molecules* **2023**, *28*, 2038. [CrossRef]
10. Kim, J.G.; Lee, J.G.; Wie, J.J. Confinement-Induced Fabrication of Liquid Crystalline Polymeric Fibers. *Molecules* **2022**, *27*, 5639. [CrossRef]
11. Zhou, Z.; Li, W.; Qian, J.; Liu, W.; Wang, Y.; Zhang, X.; Guo, Q.; Yashchyshyn, Y.; Wang, Q.; Shi, Y.; et al. Flexible Liquid Crystal Polymer Technologies from Microwave to Terahertz Frequencies. *Molecules* **2022**, *27*, 1336. [CrossRef] [PubMed]
12. Xie, C.; Yang, S.; He, R.; Liu, J.; Chen, Y.; Guo, Y.; Guo, Z.; Qiu, T.; Tuo, X. Recent Advances in Self-Assembly and Application of Para-Aramids. *Molecules* **2022**, *27*, 4413. [CrossRef] [PubMed]
13. Shimomura, Y.; Kawamura, A.; Tokita, M.; Watanabe, J.; Konishi, G. Fluorinated Poly(pentylene 4,4′-bibenzoate)s with Low Isotropization Temperatures and Unique Phase Transition Behavior. *Macromolecules* **2023**, *56*, 5152. [CrossRef]
14. Kasai, H.; Hayakawa, Y.; Akiyama, H.; Ozaki, M.; Kato, T.; Kakimoto, M. New Fabrication Approach to Develop a High Birefringence Photo-Crosslinked Film Based on a Sulfur-Containing Liquid Crystalline Molecule with Large Temperature Dependence of Birefringence. *Mol. Cryst. Liq. Cryst.* **2018**, *671*, 28–35. [CrossRef]
15. McArdle, C.B. *Side Chain Liquid Crystal Polymers*; Springer: Dordrecht, The Netherlands; Loctite Ltd.: Dublin, Ireland, 1989.
16. Ikeda, T.; Tsutsumi, O. Optical Switching and Image Storage by Means of Azobenzene Liquid-Crystal Films. *Science* **1995**, *268*, 1873. [CrossRef]
17. Hisano, K.; Aizawa, M.; Ishizu, M.; Kurata, Y.; Nakano, W.; Akamatsu, N.; Christopher, J.; Shishido, A. Scanning wave photopolymerization enables dye-free alignment patterning of liquid crystals. *Sci. Adv.* **2017**, *3*, e1701610. [CrossRef]
18. Takahashi, K.; Taguchi, D.; Kajitani, T.; Fukushima, T.; Kubo, S.; Shishido, A. Synthesis and Characterization of Side-Chain Liquid-Crystalline Block Copolymers Containing Cyano-Terminated Phenyl Benzoate Moieties. *Molecules* **2023**, *28*, 7849. [CrossRef]
19. Shigeyama, T.; Matsumoto, K.; Hisano, K.; Tsutsumi, O. Tunable Reflection through Size Polydispersity of Chiral-Nematic Liquid Crystal Polymer Particles. *Molecules* **2023**, *28*, 7779. [CrossRef]
20. Arakawa, Y.; Nakajima, S.; Kang, S.; Konishi, G.; Watanabe, J. Synthesis and evaluation of high-birefringence polymethacrylate having a diphenyl-diacetylene LC moiety in the side chain. *J. Mater. Chem.* **2012**, *22*, 14346. [CrossRef]
21. Kang, S.; Nakajima, S.; Arakawa, Y.; Tokita, M.; Watanabe, J.; Konishi, G. Highly birefringent side-chain LC polymethacrylate with a dinaphthyl-acetylene mesogenic unit. *Polym. Chem.* **2014**, *5*, 2253. [CrossRef]
22. Ma, X.; Li, Q.; Song, C.; Zhang, H.; Zhi, Y.; He, X. The effect of different photoluminescent side-chain length on the phase behaviour of chiral liquid crystal polymer. *Liq. Cryst.* **2022**, *49*, 1498–1510. [CrossRef]
23. Castro, L.D.C.; Lub, J.; Oliveira, O.N., Jr.; Schenning, A.P.H.J. Mechanochromic Displays Based on Photoswitchable Cholesteric Liquid Crystal Elastomers. *Angew. Chem. Int. Ed.* **2024**, e202413559. [CrossRef]
24. Xiong, Y.; He, P. A review on electrocatalysis for alkaline oxygen evolution reaction (OER) by Fe-based catalysts. *J. Mater. Sci.* **2023**, *58*, 2041. [CrossRef]
25. Arakawa, Y.; Kuwahara, H.; Sakajiri, K.; Kang, S.; Tokita, M.; Konishi, G. Highly birefringent polymer films from the photo-crosslinking polymerisation of bistolane-based methacrylate monomers. *Liq. Cryst.* **2015**, *42*, 1419. [CrossRef]
26. O'Neill, M.; Kelly, S.M. Ordered Materials for Organic Electronics and Photonics. *Adv. Mater.* **2010**, *23*, 566. [CrossRef]
27. Swager, T.M. Molecular Shape and Polar Order in Columnar Liquid Crystals. *Acc. Chem. Res.* **2022**, *55*, 3010. [CrossRef]
28. Hoeben, F.J.M.; Jonkheijm, P.; Meijer, E.W.; Schenning, A.P.H.J. About Supramolecular Assemblies of π-Conjugated Systems. *Chem. Rev.* **2005**, *105*, 1491. [CrossRef]

29. Sakurai, T.; Kato, K.; Shimizu, M. Side-Chain Labeling Strategy for Forming Self-Sorted Columnar Liquid Crystals from Binary Discotic Systems. *Crystals* **2023**, *13*, 1473. [CrossRef]
30. Demus, D.; Goodby, J.; Gray, G.W.; Spiess, H.W.; Vill, V. *Physical Properties of Liquid Crystals*; Wiley-VCH: Weinheim, Germany New York, NY, USA; Chichester, UK; Brisbane, Australia; Singapore; Toronto, ON, Canada, 1999.
31. Heilmeier, G.H.; Zanoni, L.A.; Barton, L.A. Guest-Host Interactions in Nematic Liquid Crystals. A New Electro-Optic Effect. *Appl Phys. Lett.* **1968**, *13*, 91. [CrossRef]
32. Heilmeier, G.H. Liquid crystal displays: An experiment in interdisciplinary research that worked. *IEEE Trans. Electron. Devices* **1976**, *23*, 780. [CrossRef]
33. O'Neill, M.; Kelly, S.M. Liquid crystals for charge transport, luminescence, and photonics. *Adv. Mater.* **2003**, *15*, 1135. [CrossRef]
34. Giménez, R.; Piñol, M.; Serrano, J.L. Luminescent liquid crystals derived from 9,10-bis(phenylethynyl)anthracene. *Chem. Mater.* **2004**, *16*, 1377. [CrossRef]
35. Fleischmann, E.K.; Zentel, R. Liquid-crystalline ordering as a concept in materials science: From semiconductors to stimuli-responsive devices. *Angew. Chem. Int. Ed.* **2013**, *52*, 8810. [CrossRef] [PubMed]
36. Iino, H.; Usui, T.; Hanna, J. Liquid crystals for organic thin-film transistors. *Nat. Commun.* **2015**, *6*, 6828. [CrossRef]
37. Padalkar, V.S.; Tsutsui, Y.; Sakurai, T.; Sakamaki, D.; Tohnai, N.; Kato, K.; Takata, M.; Akutagawa, T.; Sakai, K.; Seki, S. Optical and structural properties of ESIPT inspired HBT–fluorene molecular aggregates and liquid crystals. *J. Phys. Chem. B.* **2017**, *121*, 10407 [CrossRef]
38. Zhang, W.; Sakurai, T.; Aotani, M.; Watanabe, G.; Yoshida, H.; Padalkar, V.S.; Tsutsui, Y.; Sakamaki, D.; Ozaki, M.; Seki, S Highly fluorescent liquid crystals from excited-state intramolecular proton transfer molecules. *Adv. Opt. Mater.* **2019**, *7*, 1801349 [CrossRef]
39. Arakawa, Y.; Sasaki, S.; Igawa, K.; Tokita, M.; Konishi, G.; Tsuji, H. Birefringence and photoluminescence properties of diphenylacetylene-based liquid crystal dimers. *New J. Chem.* **2020**, *44*, 17531. [CrossRef]
40. Voskuhl, J.; Giese, M. Mesogens with aggregation-induced emission properties: Materials with a bright future. *Aggregate* **2022**, *3*, e124. [CrossRef]
41. Yamada, S.; Yoshida, K.; Uto, E.; Yoshida, K.; Sakurai, T.; Konno, T. Development of photoluminescent liquid-crystalline dimers bearing two fluorinated tolane-based luminous mesogens. *J. Mol. Liq.* **2023**, *363*, 119884. [CrossRef]
42. Yamada, S.; Konno, T. Development of donor-π-acceptor-type fluorinated tolanes as compact condensed phase luminophores and applications in photoluminescent liquid-crystalline molecules. *Chem. Rec.* **2023**, *23*, e202300094. [CrossRef]
43. Wu, X.; Niu, X.; Zhu, S.; Tian, M.; Liu, W. A novel multi-stimuli responsive fluorescence liquid crystal material with aggregationinduced emission effect. *Liq. Cryst.* **2023**, *50*, 1035. [CrossRef]
44. Sawatari, Y.; Shimomura, Y.; Takeuchi, M.; Iwai, R.; Tanaka, T.; Tsurumaki, E.; Tokita, M.; Watanabe, J.; Konishi, G. Supramolecular liquid crystals from the dimer of L-shaped molecules with tertiary amide end groups. *Aggregate* **2024**, *5*, e507. [CrossRef]
45. Iida, Y.; Shimomura, Y.; Tokita, M.; Konishi, G. Push-pull biphenyl and tolane derivatives as novel luminescent liquid crystals: Synthesis and properties. *Liq. Cryst. Early View.* **2024**, 1–14. [CrossRef]
46. Pathak, S.K.; Pradhan, B.; Gupta, M.; Pal, S.K.; Sudhakar, A.A. Liquid-Crystalline Star-Shaped Supergelator Exhibiting Aggregation-Induced Blue Light Emission. *Langmuir* **2016**, *32*, 9301. [CrossRef]
47. Demus, D. One Century Liquid Crystal Chemistry: From Vorländer's Rods to Disks, Stars and Dendrites. *Mol. Cryst. Liq. Cryst.* **2006**, *364*, 25. [CrossRef]
48. Mei, J.; Leung, N.L.; Kwok, R.T.; Lam, J.W.; Tang, B.Z. Aggregation-Induced Emission: Together We Shine, United We Soar! *Chem. Rev.* **2015**, *115*, 11718. [CrossRef]
49. Sasaki, S.; Suzuki, S.; Sameera, W.M.C.; Igawa, K.; Morokuma, K.; Konishi, G. Highly Twisted N,N-Dialkylamines as a Design Strategy to Tune Simple Aromatic Hydrocarbons as Steric Environment-Sensitive Fluorophores. *J. Am. Chem. Soc.* **2016**, *138*, 8194 [CrossRef]
50. Suzuki, S.; Sasaki, S.; Sairi, A.S.; Iwai, R.; Tang, B.Z.; Konishi, G.I. Principles of Aggregation-Induced Emission: Design of Deactivation Pathways for Advanced AIEgens and Applications. *Angew. Chem. Int. Ed.* **2020**, *59*, 9856. [CrossRef]
51. Shen, Z.; Pan, Y.; Yan, D.; Wang, D.; Tang, B.Z. AIEgen-Based Nanomaterials for Bacterial Imaging and Antimicrobial Applications: Recent Advances and Perspectives. *Molecules* **2023**, *28*, 2863. [CrossRef]
52. Luo, W.; Tan, Y.; Gui, Y.; Yan, D.; Wang, D.; Tang, B.Z. Near-Infrared-Emissive AIE Bioconjugates: Recent Advances and Perspectives. *Molecules* **2022**, *27*, 3914. [CrossRef]
53. Zhou, L.; Zheng, L.L.; Gao, M.Y.; Xu, C.J.; Ge, Y.F.; Bai, T.X.; Wen, J.; Cheng, Y.H.; Zhu, M.F. Confinement fluorescence effect of an aggregation-induced emission luminogen in crystalline polymer. *Aggregate* **2023**, *4*, e338. [CrossRef]
54. Kumari, B.; Dahiwadkar, R.; Kanvah, S. White light emission from AIE-active luminescent organic materials. *Aggregate* **2022**, *3*, e191. [CrossRef]
55. Iwai, R.; Suzuki, S.; Sasaki, S.; Sairi, A.S.; Igawa, K.; Suenobu, T.; Morokuma, K.; Konishi, G. Bridged Stilbenes: AIEgens Designed via a Simple Strategy to Control the Non-radiative Decay Pathway. *Angew. Chem. Int. Ed.* **2020**, *59*, 10566. [CrossRef]
56. Shimomura, Y.; Konishi, G. Flexible Alkylene Bridges as a Tool to Engineer Crystal Distyrylbenzene Structures Enabling Highly Fluorescent Monomeric Emission. *Chem. Eur. J.* **2022**, *28*, e202201884. [CrossRef]

57. Shimomura, Y.; Konishi, G. Push-Pull Bridged Distyrylbenzene with Highly Bright Solid-State Red-Orange Aggregation-Induced Emission. *Chem. Eur. J.* **2023**, *29*, e20231191. [CrossRef]
58. Iwai, R.; Yoshida, H.; Arakawa, Y.; Sasaki, S.; Iida, Y.; Igawa, K.; Sakurai, T.; Suzuki, S.; Tokita, M.; Watanabe, J.; et al. Near-room-temperature π-conjugated nematic liquid crystals in molecules with a flexible seven-membered ring structure. *Aggregate* **2024**, *5*, e660. [CrossRef]
59. Miyaura, N.; Suzuki, A. Palladium-Catalyzed Cross-Coupling Reactions of Organoboron Compounds. *Chem. Rev.* **1995**, *95*, 245. [CrossRef]
60. Sen, A.; Yamada, Y.M. Latest Developments on Palladium- and Nickel-Catalyzed Cross-Couplings Using Aryl Chlorides: Suzuki–Miyaura and Buchwald–Hartwig Reactions. *Synthesis* **2024**, *56*. [CrossRef]
61. Arakawa, Y.; Sasaki, Y.; Haraguchi, N.; Itsuno, S.; Tsuji, H. Synthesis, phase transitions and birefringence of novel liquid crystalline 1,4-phenylene bis(4-alkylthio benzoates) and insights into the cybotactic nematic behaviour. *Liq. Cryst.* **2017**, *45*, 821. [CrossRef]
62. Shanker, G.; Prehm, M.; Nagaraj, M.; Vij, J.K.; Tschierske, C. Development of polar order in a bent-core liquid crystal with a new sequence of two orthogonal smectic and an adjacent nematic phase. *J. Mater. Chem.* **2011**, *21*, 18711. [CrossRef]
63. Mohamed, A.; Silvio, P.; Christoph, K.; Christoph, K.; Alexey, E.; Carsten, T. Cluster phases of 4-cyanoresorcinol derived hockey-stick liquid crystals. *J. Mater. Chem. C* **2017**, *5*, 8454. [CrossRef]
64. Yuan, Y.; Li, J.; He, L.; Liu, Y.; Zhang, H. Preparation and properties of side chain liquid crystalline polymers with aggregation-induced emission enhancement characteristics. *J. Mater. Chem. C* **2018**, *6*, 7119. [CrossRef]
65. Ni, B.; Liao, J.; Chen, S.; Zhang, H. Influence of alkoxy tail length on the phase behaviors of side-chain liquid crystalline polymers without the spacer. *RSC Adv.* **2015**, *5*, 9035. [CrossRef]
66. Arakawa, Y.; Nakajima, S.; Ishige, R.; Uchimura, M.; Kang, S.; Konishi, G.; Watanabe, J. Synthesis of diphenyl-diacetylene-based nematic liquid crystals and their high birefringence properties. *J. Mater. Chem.* **2012**, *22*, 8394. [CrossRef]
67. Kang, S.; Nakajima, S.; Arakawa, Y.; Konishi, G.; Watanabe, J. Large extraordinary refractive index in highly birefringent nematic liquid crystals of dinaphthyldiacetylene-based materials. *J. Mater. Chem. C* **2013**, *1*, 4222. [CrossRef]
68. Arakawa, Y.; Kang, S.; Nakajima, S.; Sakajiri, K.; Cho, Y.; Kawauchi, S.; Watanabe, J.; Konishi, G. Diphenyltriacetylenes: Novel nematic liquid crystal materials and analysis of their nematic phase-transition and birefringence behaviours. *J. Mater. Chem. C* **2013**, *1*, 8094. [CrossRef]
69. Arakawa, Y.; Kang, S.; Tsuji, H.; Watanabe, J.; Konishi, G. The design of liquid crystalline bistolane-based materials with extremely high birefringence. *RSC Adv.* **2016**, *6*, 92845. [CrossRef]
70. Arakawa, Y.; Kang, S.; Tsuji, H.; Watanabe, J.; Konishi, G. Development of novel bistolane-based liquid crystalline molecules with an alkylsulfanyl group for highly birefringent materials. *RSC Adv.* **2016**, *6*, 16568. [CrossRef]
71. Frisch, M.J.; Trucks, G.W.; Schlegel, H.B.; Scuseria, G.E.; Robb, M.A.; Cheeseman, J.R.; Scalmani, G.; Barone, V.; Petersson, G.A.; Nakatsuji, H.; et al. *Gaussian 16, Revision C.01*; Gaussian, Inc.: Wallingford, CT, USA, 2016.
72. Kim, J.; Li, Y.; Miskiewicz, M.N.; Oh, C.; Kudenov, M.W.; Escuti, M.J. Fabrication of ideal geometric-phase holograms with arbitrary wavefronts. *Optica* **2015**, *2*, 958. [CrossRef]
73. Pi, D.; Liu, J.; Wang, Y. Review of computer-generated hologram algorithms for color dynamic holographic three-dimensional display. *Light Sci. Appl.* **2022**, *11*, 231. [CrossRef]
74. Shishido, A. Rewritable holograms based on azobenzene-containing liquid-crystalline polymers. *Polym. J.* **2010**, *42*, 525. [CrossRef]
75. Kobashi, J.; Yoshida, H.; Ozaki, M. Polychromatic Optical Vortex Generation from Patterned Cholesteric Liquid Crystals. *Phys. Rev. Lett.* **2016**, *116*, 253903. [CrossRef] [PubMed]
76. Morita, F.; Kishida, Y.; Sato, Y.; Sugiyama, H.; Abekura, M.; Nogami, J.; Toriumi, N.; Nagashima, Y.; Kinoshita, T.; Fukuhara, G.; et al. Design and enantioselective synthesis of 3D π-extended carbohelicenes for circularly polarized luminescence. *Nat. Synth.* **2024**, *3*, 774. [CrossRef]
77. Hsieh, Y.-Y.; Shyue, J.-J.; Chao, Y.-C.; Wong, K.-T.; Bassani, D.M. Chiral Binaphthalene Building Blocks for Self-Assembled Nanoscale CPL Emitters. *Molecules* **2023**, *28*, 3382. [CrossRef] [PubMed]
78. Furoida, A.; Daitani, M.; Hisano, K.; Tsutsumi, O. Aggregation-Enhanced Room-Temperature Phosphorescence from Au(I) Complexes Bearing Mesogenic Biphenylethynyl Ligands. *Molecules* **2021**, *26*, 7255. [CrossRef] [PubMed]
79. Kato, K.; Iwano, R.; Tokuda, S.; Yasuzawa, K.; Gon, M.; Ohtani, S.; Furukawa, S.; Tanaka, K.; Ogoshi, T. Circularly polarized luminescence from a common alkoxy pillar [5]arene and its co-aggregates with π-conjugated rods. *Aggregate* **2024**, *5*, e482. [CrossRef]
80. Gong, Z.L.; Li, Z.Q.; Zhong, Y.W. Circularly polarized luminescence of coordination aggregates. *Aggregate* **2022**, *3*, e177. [CrossRef]
81. Yan, H.; He, Y.; Wang, D.; Han, T.; Tang, B.Z. Aggregation-induced emission polymer systems with circularly polarized luminescence. *Aggregate* **2023**, *4*, e331. [CrossRef]
82. Uchimura, M.; Watanabe, Y.; Araoka, F.; Watanabe, J.; Takezoe, H.; Konishi, G. Development of Laser Dyes to Realize Low Threshold in Dye-Doped Cholesteric Liquid Crystal Lasers. *Adv. Mater.* **2010**, *22*, 4473. [CrossRef]
83. Wan, S.-P.; Lu, H.-Y.; Li, M.; Chen, C.-F. Advances in circularly polarized luminescent materials based on axially chiral compounds. *J. Photochem. Photobiol. C* **2022**, *50*, 100500. [CrossRef]

84. Hasegawa, Y.; Kitagawa, Y. Luminescent lanthanide coordination polymers with transformative energy transfer processes for physical and chemical sensing applications. *J. Photochem. Photobiol. C* **2022**, *51*, 100485. [CrossRef]
85. Cho, S.; Takahashi, M.; Fukuda, J.; Yoshida, H.; Ozaki, M. Directed self-assembly of soft 3D photonic crystals for holograms with omnidirectional circular-polarization selectivity. *Commun. Mater.* **2021**, *2*, 39. [CrossRef]

MDPI AG
Grosspeteranlage 5
4052 Basel
Switzerland
Tel.: +41 61 683 77 34

Molecules Editorial Office
E-mail: molecules@mdpi.com
www.mdpi.com/journal/molecules

www.ingramcontent.com/pod-product-compliance
Lightning Source LLC
Chambersburg PA
CBHW062031220926
51850CB00010B/2514